Transfer Function
Techniques and
Fault Location

MECHANICAL ENGINEERING RESEARCH STUDIES

ENGINEERING DYNAMICS SERIES

Series Editor: **Dr J. B. Roberts,** *University of Sussex, England*

1. Feedback Design of Systems with Significant Uncertainty
   M. J. Ashworth

2. Modal Testing: Theory and Practice
   D. J. Ewins

3. Transfer Function Techniques and Fault Location
   J. Hywel Williams

4. Parametric Random Vibration
   R. A. Ibrahim

Transfer Function Techniques and Fault Location

J. Hywel Williams

UWIST, Cardiff, Wales

RESEARCH STUDIES PRESS LTD.
Letchworth, Hertfordshire, England
JOHN WILEY & SONS INC.
New York · Chichester · Toronto · Brisbane · Singapore

RESEARCH STUDIES PRESS LTD.
58B Station Road, Letchworth, Herts. SG6 3BE, England

Marketing and Distribution:

Australia, New Zealand, South-east Asia:
Jacaranda-Wiley Ltd., Jacaranda Press
JOHN WILEY & SONS INC.
GPO Box 859, Brisbane, Queensland 4001, Australia

Canada:
JOHN WILEY & SONS CANADA LIMITED
22 Worcester Road, Rexdale, Ontario, Canada

Europe, Africa:
JOHN WILEY & SONS LIMITED
Baffins Lane, Chichester, West Sussex, England

North and South America and the rest of the world:
JOHN WILEY & SONS INC.
605 Third Avenue, New York, NY 10158, USA

Library of Congress Cataloging in Publication Data:

Williams, J. Hywel, 1934–
    Transfer function techniques and fault location.

    (Mechanical engineering research studies)
(Engineering dynamics series: No. 3)
    Bibliography: p.
    Includes index.
    1. Fault location (Engineering)   2. Transfer
functions.   I. Title.   II. Series.   III. Series:
Engineering dynamics series; 3.
TA169.6.W55   1985   624.1'51   85-9443
ISBN 0 471 90805 3 (Wiley)

British Library Cataloguing in Publication Data:

Williams, J. Hywel
    Transfer function techniques and fault location.
    —(Mechanical engineering research studies)—
    (Engineering dynamics series; 3)
    1. System failures (Engineering)   2. Dynamics
    I. Title   II. Series
    620.7'3'   TA169.5

    ISBN 0 86380 031 9
    ISBN 0 471 90805 3 (Wiley)

    ISBN 0 86380 031 9 (Research Studies Press Ltd.)
    ISBN 0 471 90805 3 (John Wiley & Sons Inc.)

Printed in Great Britain

Preface

This monograph does not attempt to give a comprehensive
review of fault location methods using statistically
derived pattern recognition methods. The aim is to
produce a practical methodology and background to enable
the test engineer to establish a diagnostic system based
upon his own experience of the system. The contents are
based upon the research and development carried out in
the Dynamic Analysis Group at U.W.I.S.T. This group was
established in 1967 by Denis Towill. I have been
fortunate in having him as a mentor and friend. He has
had a significant influence on so many careers, such as
those of David Lamb, Peter Payne, H. Sriyananda, K.C.
Varghese. My scratchings are an attempt to repay some of
the debt.

J. Hywel Williams
Dynamic Analysis Group
U.W.I.S.T.
CARDIFF

To THEMUNA ANN

Contents

Contents

CHAPTER 1
Introduction—
A Review

1.1. Introduction

Over recent years there has been much effort in
developing testing and diagnostic methods for digital
circuits. With the growth in digital technology this is
to be expected. However, the problems of fault location
in analogue systems are still of paramount importance.
By analogue systems we mean more than electronic or
electro-mechanical configurations. In fact, systems that
have definable or measurable input/output relationships
which are not of a binary nature can be considered to
pose similar problems to the diagnostician.

1.2. System Checkout and Fault Location

Fault location is based upon the assumption that the
failure of an assembly to perform properly is caused by

one or more faulty parts of the assembly. The purpose of
a diagnostic scheme is to isolate a fault to the smallest
replaceable item. The fundamental philosophy is that the
fault to be detected causes a deviation in one or more of
the output responses of sufficient magnitude to move the
operating point outside the region of normal operation.

System testing has two phases:-

   (i)   checkout
   (ii)   fault location

The checkout phase is of a binary nature i.e. pass/fail;
go/no go. This is shown diagrammatically in an arbitary
two dimensional measurement space, in Figure 1.1.

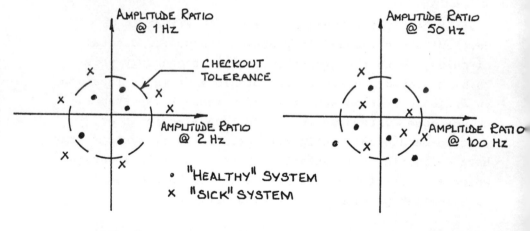

(i) Measurements with   (ii) Measurements with
    good discrimination   poor discrimination

FIG 1.1. "HEALTHY" AND "SICK" SYSTEMS IN AN
         ARBITARY TWO DIMENSIONAL MEASUREMENT SPACE

This simple representation indicates one of the
fundamental problems encountered in fault location;
namely that the robustness of the technique depends on a
good choice of test. It is usual that the pass/fail
decision logic is based upon the tolerance assigned to
the system's response. Applications of sensitivity
analysis (SHOOMAN 1968), Monte Carlo simulation and
updating tolerance limits as field data is accumulated
(PAYNE, TOWILL, BAKER 1970) are all suitable for this
purpose.

The second phase, of fault location, may include
additional tests but is mainly concerned with processing
the test data so as to identify the cause of failure.
Fault diagnostic methods can be categorised into two
classes. Those that estimate the "value" of the fault
component and those that only locate the fault; the
actual value of the component is often of no interest.
The schematic of the testing philosophy is shown in
Figure 1.2.

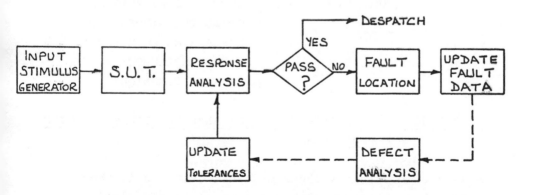

FIGURE 1.2. TESTING SCHEMATIC

4

1.3. Fault Diagnostic Methods

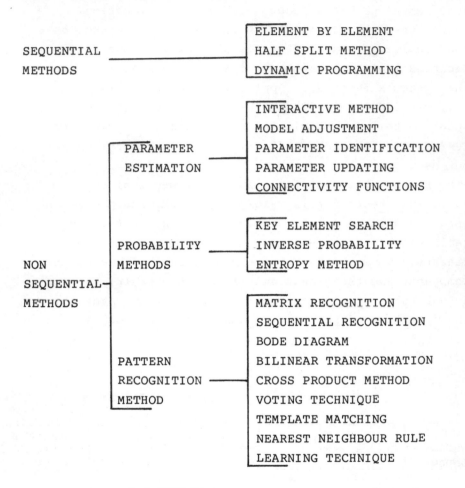

SEQUENTIAL METHODS
- ELEMENT BY ELEMENT
- HALF SPLIT METHOD
- DYNAMIC PROGRAMMING

NON SEQUENTIAL METHODS

PARAMETER ESTIMATION
- INTERACTIVE METHOD
- MODEL ADJUSTMENT
- PARAMETER IDENTIFICATION
- PARAMETER UPDATING
- CONNECTIVITY FUNCTIONS

PROBABILITY METHODS
- KEY ELEMENT SEARCH
- INVERSE PROBABILITY
- ENTROPY METHOD

PATTERN RECOGNITION METHOD
- MATRIX RECOGNITION
- SEQUENTIAL RECOGNITION
- BODE DIAGRAM
- BILINEAR TRANSFORMATION
- CROSS PRODUCT METHOD
- VOTING TECHNIQUE
- TEMPLATE MATCHING
- NEAREST NEIGHBOUR RULE
- LEARNING TECHNIQUE

FIGURE 1.3. CLASSIFICATION OF ANALOGUE FAULT LOCATION METHODS

A list of possible methods of fault location is shown in Figure 1.3. Primarily they can be classified into sequential and non-sequential methods (BAINTON, ROWLANDS, TOWILL, WILLIAMS 1978).

1.3.1. Sequential Methods
1.3.1.1 Element by element testing (SHIELDS 1976)

 Fault location is achieved by proceeding logically
through the test procedure and finding the failed
elements by a process of elimination. It is assumed that
the various elements of the system have the same
probability of failure and all testing costs are the
same. The order of the tests is thus unimportant and a
random strategy is often used. When the number of
elements becomes large this method becomes cumbersome.

1.3.1.2. Half Split Method (SHIELDS 1976)

 If a system consists of N testable sub units, N/2 are
tested. Whichever group contains the fault, that group
is further sub-divided until the fault is isolated. This
method is faster than the element by element method
assuming that all faults are equally likely.

1.3.1.3. Dynamic Programming (GLUSS 1962)

 Methods based on dynamic programming have been used
which minimise the ratio of probabilities of failure to
testing costs. This is particularly applicable when the
system has built-in redundancy.

1.3.2. Non-Sequential Methods
1.3.2.1. Pattern Recognition Methods

 The general approach can be stated as "given a set of
responses made on the system under test, classify the
system into one of several predetermined categories in

which it is known the system belongs". This implies that
responses under fault conditions are stored and the
response of the system under test is compared with the
stored responses to identify the fault. It is unlikely
that there will be an exact match so criteria must be
established to define a "best match" within the stored
set of responses.Most techniques in this category require
a fault dictionary or recognition matrix to be formed
which is the matrix of simulated or accumulated
"a priori" failure information of likely faults on the
system under test. Each representation in the matrix is
a deviation vector evaluated as the deviation in response
from the nominal due to a particular fault at the
pre-specified measurements or features and is represented
as

$$\underline{x} = \{x_1 \ x_2 \ x_3 \ \ldots\ldots\ldots x_m\}^T \qquad\qquad 1.1$$

Such vectors for all likely faults form the matrix

$$R = \{x_{ij}\} \qquad\qquad 1.2$$
$$i=1,2,\ldots M \text{ measurements}$$
$$j=1,2,\ldots N \text{ fault cases}$$

The recognition matrix can be set up by evaluating the
deviation vectors x by simulation of the mathematical
model of the SUT or by physically changing parameter
values to introduce faults or by using classic
sensitivity functions (SRIYANANDA, TOWILL 1973). Various
coding methods of the matrix entries are used prior to
use in a diagnostic scheme (VARGHESE, WILLIAMS, TOWILL
1978). Selection of the measurements or features is
complex and will be considered in detail later.

1.3.2.2. Matrix Recognition (GARCIA 1971)

A number of test points are chosen based on which the
regions of normal operation and that of failure are
defined. Failure classification is achieved by varying
the number of test points. A matrix is formed in such a
way that its product with the input vector gives the
output vector and the highest component of the output
vector gives the region which is out of operation. The
component corresponding to this region is the faulty
component.

1.3.2.3. Sequential Recognition Technique
          (BERKOWITZ, KRISHNASWANY 1963)

This is similar to the method of matrix recognition but
it is only necessary to check for each region if the
boundaries are met or not by the measurements.

1.3.2.4. Bode Diagram Technique (GARCIA 1971)

This technique involves the setting up of a fault
dictionary which is the list of all possible component
value variations with the associated variations in the
gain at selected frequencies. The variations in gain are
coded "+","-" or "0" based on whether the gain is above,
below or within tolerance limits. A fault in the SUT is
diagnosed by comparing the pattern vector with the stored
pattern. Test frequencies are selected by picking one
below the lowest non zero break point, one above the
highest break point and one mid way (on a log scale)
between adjacent break points. In addition one is added
at each complex break point.

1.3.2.5. Voting Technique (SRIYANANDA, TOWILL 1972)

The voting method quantitises the deviation in response
from nominal into +1,-1 or 0 as in the Bode diagram
technique. Thus the recognition matrix only has entries
+1,-1 or 0. The response deviation vector of the SUT is
quantitised in the same manner. The features can be
specific time delays of the system's cross correlation
function at multiples of the pseudo-noise stimulus clock
period, discrete sine wave frequencies, sampled points of
the step response etc.
The deviation vector of the SUT is compared with the
stored patterns and votes cast (i) in favour of the fault
(ii) against the fault or (iii) no votes are cast.
Several forms of logic exist to determine the way in
which votes are cast. When all the votes have been cast
the faults are ranked according to the net votes cast.
The fault with the most votes being the most likely fault
etc. Methods of feature selection are considered later.

1.3.2.6. Template Matching (STAHL, MAENPAA 1969)

In this method the response at each data point is coded
according to the size of the deviation from nominal. For
example, if phase deviations from nominal are coded into
bands, then measurements of say (4°) (-8°) (-35°) and
(-12°) may be coded as 1, 1, 4, 2. The coded result is
then compared with similar coded patterns of known faults
and equivalence indicates the fault. It is unlikely that
an exact match will be found in analogue systems of
reasonable production variation.

1.3.2.7. Cross-Product Method (SRIYANANDA,TOWILL 1973)

 The principle involved is that the greatest correlation
will exist between the measured deviation and the stored
deviation for the actual fault. So the stored vector
which has the highest cross-product with the vector of
the SUT indicates the most likely fault.

1.3.2.8. Bi-Linear Transformation (MARTENS, DYCK 1972)

 For a given parameter p the transformation

$$T(p) = \frac{Ap + B}{Cp + D}$$

1.3

where A,B,C and D are complex constants is known as a
bi-linear transformation. It transforms straight lines in
the p domain to circles in the T domain. If a system's
transfer function can be written in the form

$$T(s,p) = \frac{A(s)p + B(s)}{C(s)p + D(s)}$$

1.4

where s is the Laplace operator, then substitution of jw
for s yields the system's frequency response and (1.4)
represents a bi-linear transformation. Thus at a given
frequency, circular loci can be drawn in the T domain for
each parameter p. The loci will intersect at the nominal
point for the system. Under single fault conditions the
response of the SUT will lie on one of the loci, so
identifying the fault. However, the natural deviations in
the non-faulty parameters make the probability of lying
on a locus unlikely. Further, as the number of parameters
is large the method is cumbersome.

1.3.2.9. Nearest Neighbour Rule (TOWILL, WILLIAMS 1977)

The nearest neighbour rule classification assigns a test
vector of the SUT to one of the many possible fault
classes by selecting the "nearest" stored vector. The
simplest metric to use is the distance (N dimensions)
between vectors. Thus x (one of the stored vectors) is
the nearest neighbour to y (test vector of the SUT) if
the distance $D(x_j,y)$ is the minimum where

$$D(x_j,y) = \min_i\{D(x_i,y)\} \qquad\qquad 1.5$$

$$\text{where } x_i \epsilon\{x_1\ x_2\ x_3\ \ldots\ldots\ldots x_N\}$$

$$\text{and } D(x_j,y) = \{\Sigma(x_{kj} - y_k)^2\}^{\frac{1}{2}} \qquad\qquad 1.6$$

Whilst the approach is simple and attractive care has to
be taken in normalising the data to account for wide
parameter variations and keeping the recognition matrix
of minimum dimension to facilitate efficient data
processing. The technique will be considered later in
greater detail.

1.3.2.10. Fuzzy Sets (BEDROSIAN 1978)

In any realistic system the effects of measurement noise
and non-faulty component value variation create a
"fuzziness" in the boundaries of the various fault
classes. A weighting function having values between 0 and
1 is used to grade the membership of an element in a
set. The nearer to 1 then the degree of belonging of the
element increases.

1.3.3. Probability Methods

1.3.3.1. Key Element Search Method (BUCHSBAUN, DUNNING, HANNOM, MOTH 1964)

The test data is used to calculate a fault index which minimises the difference between the observed deviations from nominal and the deviations predicted from a knowledge of the sensitivity functions. The component value that minimises the index is considered to be the most likely fault. This approach will be amplified later.

1.3.3.2.Inverse Probability (BROWN,McALLISTER,PERRY 1963)

The differences between the nominal values and the measured values of the response of the SUT are used to estimate the probability of failure of each individual component of the system. The component having the highest probability is naturally the most likely fault.

1.3.3.3. Entropy Method (GOOD 1968)

Entropy is a statistical measure of uncertainty and is defined as

$$H = p_i \log(p_i) \qquad\qquad 1.7$$

Where p is the probability of the ith item failure. H measures the interset dispersion for a set of units of known failure probabilities and is used for probabilistic decision making in the diagnosis.

1.3.4. Parameter Estimation Methods

1.3.4.1. Interactive Method (BERKOWITZ 1962)

This method requires that the system's topology and
nominal component values are known. The variability of
parameter values is also assumed known. The test data is
used in a maximum likelihood estimation routine to
estimate the component values. Comparing these estimates
with the nominal values indicates the faulty components.
Probabilistic procedures require large amounts of test
data to produce stable estimates.

1.3.4.2. Model Adjusting Technique (TOWILL, BAKER 1970)

This approach is an extension of a system design
technique. A mathematical model of the system is first
established; typically the system's transfer function.
Experimental data are then used in a least squares curve
fitting routine to estimate the system's transfer
function coefficents.The relationships between these
coefficients and the component values then indicate the
faulty component.

1.3.4.3. Parameter Updating (SRIYANANDA 1972)

This approach is appropriate to process systems where
data is used on line to update parameter values. The
Kalman filter and potential functions are typical
techniques in this area.

The selection of the "best" features or measurements
(i.e. test frequencies or time delays etc.) for fault
location is of the utmost importance as the diagnostic
power and manageability of the procedure depends very
much on the feature selection. The diagnostician must
know or measure the dynamic characteristics of the system
and the effects of component value changes on the
system's response. Many methods of feature selection have
been developed and these will be examined later.

1.4. Conclusion

I have attempted, in this chapter, to give a general
review of diagnostic methods and approaches currently
available. Whilst I have my favourites, I will consider
examples of the techniques in detail so that the reader
can select for himself the approach that is appropriate
to his problems. It is emphasised that nothing is really
sacrosanct. The worker must adjust his tools and try
modifications as he sees fit. I hope I can point him in
the right direction.

1.5.   References

M.L.Shooman   "Probabilistic Reliability: An
                     Engineering Approach"
                     N.Y. McGraw Hill, 1968

P.A.Payne,D.R.Towill,K.J.Baker
                     "Predicting Servomechanism Dynamic
                     Response from Limited Production
                     Test Data"
                     Radio & Electronic Engineer,Vol 40
                     No.6 1970

D.J.Bainton,A.R.Rowlands,D.R.Towill,J.H.Williams
                     "Software Development for Analog
                     Avionic System Fault Diagnosis at
                     an ATE Test Station"
                     Automatic Testing '78, Paris 1978

S.Shields   "A Review of Fault Detection Methods
                     for Large Systems"
                     Radio & Electronic Engineer Vol.46,
                     No. 6 1976

B.Gluss   "An Optimum Policy for Detecting a
                     Fault in a Complex System"
                     Operations Research Vol. 7,No. 4 1959

H.Sriyananda,D.R.Towill
                     "Fault Diagnosis Using Time Domain
                     Measurements"
                     Radio & Electronic Engineer Vol.43,
                     No.6 1973.

R.C.Garcia   "Fault Isolation Computer Methods"
                      NASA Contract Report, CR-1759,
                      Computer Sciences Corp. Alabama 1971

R.S.Berkowitz,P.S.Krishraswany
                      "Computer Techniques for Solving
                      Electric Circuits for Fault Location"
                      IEEE Trans on Aerospace Support Conf.
                      Procedure Vol.AS-1,No.2 1963

H.Sriyananda,D.R.Towill
                      "Fault Diagnosis via Automatic
                      Dynamic Testing-A Voting Technique"
                      The Automation of Testing, IEE
                      Conf. Publication No. 91 1972

W.J.Stahl,J.H.Maenpaa
                      "Development of Advanced Dynamic
                      Fault Diagnostic Techniques"
                      Scully Int. Inc., AD-814457, 1967

G.O.Martens,J.D.Dyck
                      "Fault Identification in Electronic
                      Circuits with the aid of Bi-linear
                      Transformation"
                      IEEE Trans.Rel. Vol.R-21,No.2 1972

D.R.Towill,J.H.Williams
                      "Fault Diagnosis for a Complex
                      Electro-hydraulic System"
                      Proc.Int.Conf.on Technical
                      Diagnostics. Prague 1977

16

S.D .Bedrosian   "Analog ATPG: A Response to Users
                        Needs"
                        Autotestcon '78, 1978

L.Buchsbaum,M.Dunning,T.J.B.Hannon,L.Moth
                        "Investigation of Fault Diagnosis
                        by Computational Methods"
                        Pennington Rand, AD-601204, 1964

F.D.Brown,N.F.McAllister,R.D.Perry
                        "An Application of Inverse
                        Probability to Fault Isolation"
                        IRE Trans.Military Elect. 1963

I.J.Good   "Some Statistical Methods in
                        Machine Intelligence"
                        Research,Virginia J. of Sci.,19,1968

R.S.Berkowitz   "Conditions for Network Element
                        Value Solvability"
                        IRE Trans. Circuit Theory 1962

D.R.Towill,K.J.Baker
                        "Development of Analogue Modelling
                        Strategies for Dynamic Production
                        Testing"
                        Int. J. Prod. Res. Vol.8,No.6,1972

CHAPTER 2
Dynamic Testing

2.1. Introduction

 A dynamic system is characterised or modelled by a
differential equation or sets of such equations. The
solution of differential equations both analytically and
by numerical methods is a central topic in mathematics.
The constraints due to the fact that the differential
equations represent realisable systems enable linear
time invariant systems to be parsimoniously modelled in
the Laplace domain.
 Consider the arrangement below. It shows a dynamic
system responding to an input (time dependent) x(t) with
an output, also time dependent, y(t).

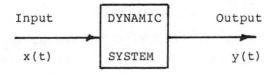

In general the differential equation relating input to output will be of the form

$$a_n \frac{d^n y(t)}{dt^n} + a_{n-1}\frac{d^{n-1}y(t)}{dt^{n-1}} + a_{n-2}\frac{d^{n-2}y(t)}{dt^{n-2}} + \ldots \ldots + a_1\frac{dy(t)}{dt} + a_0 y(t) =$$

$$b_m \frac{d^m x(t)}{dt^m} + b_{m-1}\frac{d^{m-1}x(t)}{dt^{m-1}} + b_{m-2}\frac{d^{m-2}x(t)}{dt^{m-2}} + \ldots \ldots + b_1\frac{dx(t)}{dt} + b_0 x(t)$$

2.1

Taking Laplace transforms and assuming initial conditions, including all deviations, are zero then:

$$(a_n s^n + a_{n-1}s^{n-1} + a_{n-2}s^{n-2} + \ldots \ldots + a_1 s + a_0)y(s) =$$

$$(b_m s^m + b_{m-1}s^{m-1} + b_{m-2}s^{m-2} + \ldots \ldots + b_1 s + b_0)x(s)$$

2.2

Then the relationship between input and output is given by

$$\frac{y(s)}{x(s)} = \frac{\sum\limits_{i=0}^{m} b_i s^i}{\sum\limits_{i=0}^{n} a_i s^i}$$

2.3

This ratio is called the transfer function of the system. The main advantage in using the Laplace domain is that the analysis of systems or the calculation of responses for given inputs relies on relatively simple algebraic manipulations.

2.2.   Dynamic Testing

2.2.1.   Introduction

The dynamic response of the SUT is defined as the behaviour of the system when stimulated by a time varying input such as the unit step or one of the alternative

signals considered later. Consequently, a dynamic test
is any test that yields information on the dynamic
response of the SUT even if the data yielded do not
completely describe the dynamic behaviour of the system.
Thus if only the final value of the response is measured
the test would be classified as a static test, whilst
sampling the transient behaviour would constitute a
dynamic test.

Dynamic testing offers advantages over static or steady
state testing other than when considering fault
location. Some examples are considered.

2.2.2.   User Confidence

Many systems are designed to respond to real time varying
input signals, so that a high degree of correlation
exists between operational success and the setting of
suitable dynamic performance specifications.

2.2.3.   Spares Inventory Reduction

If only static performance tests are used to establish
the status of a system which has to meet a dynamic
operational capability, it is found that static
tolerances have to be made excessively tight. So many
"healthy" systems are wrongly categorised as "sick",
resulting in the setting up of an over large spares
inventory in order to provide a reasonable level of
system availability.

2.2.4.   Increased System Reliability

Dynamic tests can be designed to reduce the need for
intermediate access points used to inject or monitor
signals needed in static test schemes. It is argued that
the reduction of access points increases the systems'
reliability.

2.2.5.   Performance Matching

Improved matching during selective assembly is possible
if dynamic performance data are made available for
individual sub-systems e.g. using frequency plots to
match the correct value of tuning resistor to an amplifer
(R-C coupled). Also reasonable tolerances must be
permitted in component values if systems are to be
produced at reasonable costs. It is often then necessary
to "fine tune" a complex system on final assembly.
Dynamic testing greatly assists this tuning procedure.

2.2.6.   Design Proving

At the system development stage, extensive dynamic
testing is required to verify the various transfer
function models used in the design stage. This will
require the use of identification and parameter
estimation routines with the dynamic test data as input.
It is also necessary to establish the likely variation in
these mathematical models which may be expected during
normal operation.

It is this last reason for dynamic testing that has
received most attention in the literature. However, the

other reasons given are of equal importance, particularly
at the high volume end of the market with its emphasis on
fast turn round to achieve high productivity from
production and maintanance test stations.

2.3.   Dynamic Tests
2.3.1. Introduction

For a linear system, a transfer function of the form

$$H(s) = \frac{\sum_{o}^{m} b_i s^i}{\sum_{o}^{n} a_i s^i}$$

enables the output of the system to be determined for any
given input. If the input stimulus x(t) has the Laplace
transform X(s) then the output response y(t) is given by

$$y(t) = L^{-1}\{H(s)X(s)\} \qquad\qquad 2.4$$

(where L^{-1} represents the inverse Laplace transform)

2.3.2. Step Response

The step response is commonly used to test dynamic
systems. It is simple to generate and its response is
intuitively easy to understand.

The Laplace transform of the unit step is 1/s, so the
system's unit step response is given by

$$y(t) = L^{-1}\{\frac{H(s)}{s}\} \qquad\qquad 2.5$$

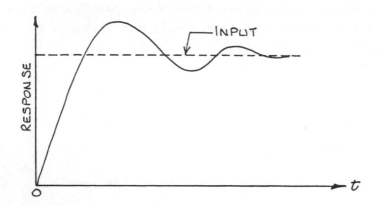

FIG. 2.1. Typical Step Response

Manufactured systems always have inherent uncertainty in their responses due to natural distributions of parameter values of their components. Figure 2.2 shows a typical go/no go mask used with a system's step response.

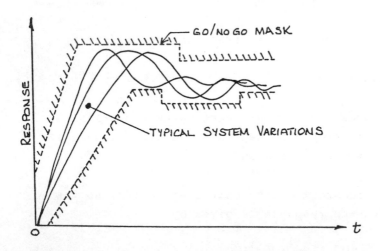

FIG. 2.2. Step Response Go/No Go Mask

Figure 2.3 shows the usual measures considered when using
the step response. The importance of any individual
criterion depends on the SUT application

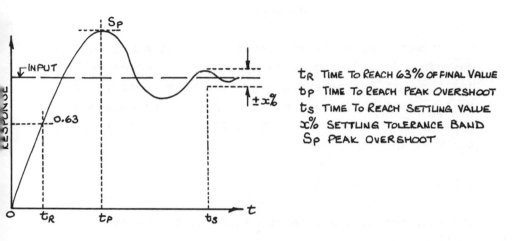

t_R TIME TO REACH 63% OF FINAL VALUE
t_P TIME TO REACH PEAK OVERSHOOT
t_S TIME TO REACH SETTLING VALUE
x% SETTLING TOLERANCE BAND
S_P PEAK OVERSHOOT

FIG. 2.3. Step Response Characteristics

2.3.3.   Measurement Noise

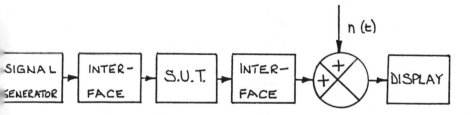

FIG 2.4,   Test Configuration with Added Noise

In a practical test situation one can not completely
eliminate the effect of measurement noise. It is
convenient to consider the noise as emanating from a
source that is independent from the SUT, see Figure 2.4.

24

It is usual to represent the noise in a probabilistic sense by estimating its statistical measures. In this way the total test configuration can be mathematically modelled and the noise rejection properties of the test techniques evaluated and compared.

The presence of noise increases the uncertainty in correctly determining go/no go states. A common method of dealing with non-deterministic random noise is to average over a number of trials so as to produce an unbiased estimate of the response. This repetition results in an increase in measurement time.

2.3.4. Impulse Testing

The fundamental test, in the mathematical sense, is the Impulse test.

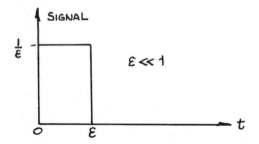

FIG 2.5. The Impulse as a Limiting Case of the Step Input

Consider the pulse shown in Figure 2.5. The area of the pulse is unity. If we now let $\varepsilon \rightarrow 0$, in the limit we obtain the unit impulse. Practically, of course, we can only generate a crude approximation to the true impulse.

However, there is a correlation technique which is considered later that indirectly measures the impulse response of a system.

As the Laplace transform of the unit impulse is 1, the Impulse response of a system is simply the Laplace inverse of the system's transfer function $L^{-1}(H(s))$. Mathematically, knowledge of the impulse response enables the response to any given time domain input to be determined, in the time domain, using the convolution integral.

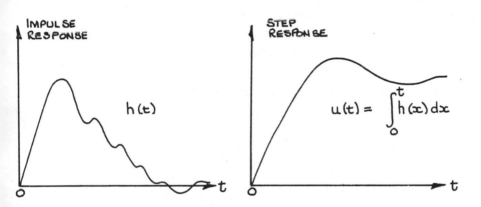

FIG 2.6.   Secondary Resonance Attenuation due to Integration

Although, in theory, the impulse and the step responses contain the same information on system performance, in practice the extraction of the information can be made difficult by an unsatisfactory choice of stimulus. It is often overlooked that the impulse test produces an amplification of the SUT's secondary resonances compared with the step response, see Figure 2.6. It should be

remembered that the unit impulse input is the time
derivative of the unit step input. Alternatively, any
secondary resonance present in the impulse response is
attenuated in the step response by the integrating
(smoothing) effect.

2.4.   Indirect Impulse Testing
2.4.1.   Introduction

The convolution integral relates the system's output to
the system's stimulus via the system's impulse response.
If fo(t), fi(t) and h(t) represent the system's output,
input and impulse response then;

$$f_o(t) \;=\; \int_{-\infty}^{\infty} h(x)f_i(t-x)\,dx \qquad\qquad 2.6$$

Now the cross-correlation function between the two
signals fi(t) and fo(t) is defined as

$$\phi_{io}(\tau) \;=\; \mathop{L}_{T\to\infty} \frac{1}{2T} \int_{-T}^{T} f_i(t-\tau)f_o(\tau)\,dt \qquad\qquad 2.7$$

Substituting for fo(t) gives

$$\phi_{io}(\tau) \;=\; \mathop{L}_{T\to\infty} \frac{1}{2T} \int_{-T}^{T} f_i(t-\tau) \int_{-\infty}^{\infty} h(x)f_i(t-x)\,dx\,dt \qquad\qquad 2.8$$

Changing the order of integration and averaging gives

$$\phi_{io}(\tau) \;=\; \int_{-\infty}^{\infty} h(x)\,dx \mathop{L}_{T\to\infty} \frac{1}{2T} \int_{-T}^{T} f_i(t-x)f_i(t-\tau)\,dt \qquad\qquad 2.9$$

If we now define the autocorrelation as

$$\phi_{ii}(\tau) \;=\; \mathop{L}_{T\to\infty} \frac{1}{2T} \int_{-T}^{T} f_i(\tau-x)f_i(x)\,dx \qquad\qquad 2.10$$

Then (2.9) can be written as

$$\phi_{io}(\tau) = \int_{-\infty}^{\infty} h(x)\phi_{ii}(\tau-x)dx \qquad 2.11$$

We can think of this as representing the input/output relationship as shown

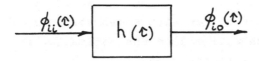

Now

$$\int_{-\infty}^{\infty} \psi(x)\delta(x-a)dx = \psi(a) \qquad 2.12$$

Where $\delta(x)$ is the Dirac delta function. So if $\phi_{ii}(\tau)$ was an impulse then $\phi_{io}(\tau)$ would be the system's impulse response.

Equation (2.11) is called the Wiener-Hopf equation. The integration limits are $\pm\infty$; however, in any practical situation we assume that time starts at zero and that the system's impulse response will be effectively zero after some finite time, T say. Thus (2.11) can be practically written as

$$\phi_{io}(\tau) = \int_{o}^{T} h(x)\phi_{ii}(\tau-x)dx \qquad 2.13$$

The emphasis thus shifts to finding an input stimulus whose autocorrelation is impulse like.

2.4.2. PNS Characteristics
2.4.2.1. Introduction

White noise is one test stimulus which has a unit impulse
autocorrelation function. Unfortunately an infinite
number of experiments is required for satisfactory
estimates of the impulse response. We are forced to use
a pseudo-random sequence with known statistical
properties which will enable the estimation to be
accomplished in a finite time.

2.4.2.2. PRBS (Hughes, Norton 1962)

Two level sequences (pseudo-random binary sequences) are
particularly attractive since they are easily generated
by shift registers incorporating the necessary feedback
and operating in modulo -2 arithmetic. Figure 2.7 shows
a typical PRBS sequence with the corresponding
autocorrelation function.

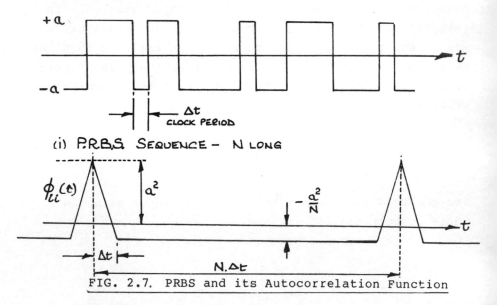

FIG. 2.7. PRBS and its Autocorrelation Function

The mechanisation of the test is shown in Figure 2.8.

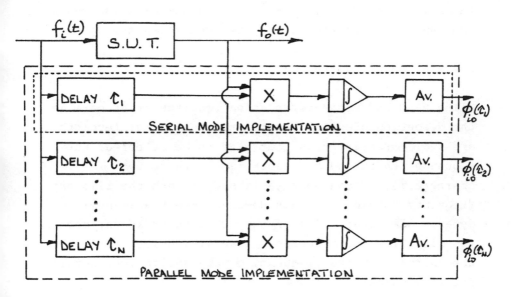

FIG. 2.8. Mechanisation of PRBS Testing

The method of test can be either serial mode or parallel
mode depending on the number of delays provided or the
provision of intermediate storage prior to computation.
Microprocessors, are of course, an ideal vehicle for PRBS
generation and signal processing.

If we examine the autocorrelation function it can be seen
that it is not a perfect impulse but is represented as:

$$\phi_{ii}(\tau) = \{\tfrac{N+1}{N}\}a^2\Delta t\delta(\tau) \quad - \quad \frac{a^2}{N} \qquad\qquad 2.14$$

The "spike" repeats itself with a periodicity of
$N\Delta t$, where $N=2^R -1$; R being the length of the shift
register. The dc offset, $-a^2/N$, reduces as N increases
and can be eliminated if reversed sequences are used.

2.4.2.3. Matching PRBS to SUT

Using PRBS and cross-correlation requires great care in
the selection of PRBS. Apart from the signal level there
are two characteristics that have to be selected; the
sequence length and the clock rate. These two
characteristics can be established in both the time and
frequency domain. In both domains, the fundamental
dynamic characteristics of the SUT must be known; such as
settling time and resonances. We will briefly describe
the approach in the frequency domain and in more detail
the time domain.

The PRB has a line spectrum in the frequency domain whose
amplitude falls off to zero at the clock frequency of the
sequence. Therefore one has to ensure that the amplitude
response of the SUT has dropped effectively to zero well
before the clock frequency. In addition the PRBS
characteristics are adjusted so that at least two
spectrum lines fall within any SUT resonances present.
In effect the PRBS test can be considered as a frequency
domain test where frequencies (line spectrum) are
inputted in parallel. A correlation technique is
normally used to extract the amplitude and phase
responses either in serial or parallel modes. Before we
consider the design of the PRBS in the time domain we
examine possible corrections to the cross-correlation
function that are required.

By considering the ramp response of the SUT the
expression given in (2.14) can be more accurately
represented to take into account waveform dependent
factors (Hughes, Noton 1962).

For $\tau > \Delta t$

$$\phi_{io}(\tau) = a^2\frac{(N+1)}{N}\Delta t\{h(\tau) + \frac{(\Delta t)^2}{12}h^{11}(\tau) + \frac{(\Delta t)^4}{60}h^{iv}(\tau) + \cdots\} \quad 2.15$$

It is usual that these derivatives are not large and of
course are discounted by powers of Δt.

For $\tau < \Delta t$

$$\phi_{io}(\tau) = a^2\frac{(N+1)}{N}\Delta t\{h(\tau) + \frac{\Delta t}{6}h^1(\tau) + \cdots\} \quad 2.16$$

Since $h'(o)$ is often relatively large, significant
errors can arise in estimating $h_{io}(o)$ from $\phi_{io}(o)$. It is
better to calculate $h(o)$ from the expression

$$h(o) = \frac{N}{a^2(N+1)\Delta t}\{3\phi_{io}(\Delta t) - 3\phi_{io}(2\Delta t) + \phi_{io}(3\Delta t)\} \quad 2.17$$

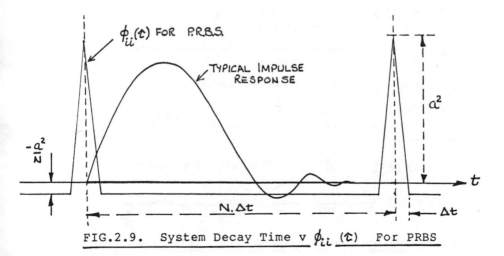

FIG.2.9. System Decay Time v $\phi_{ii}(\tau)$ For PRBS

32

Even with a narrow (small Δt) autocorrelation function
the initial value will be in error and (2.17) should be
used. A fundamental requirement is that NΔt is longer
than the SUT's delay time, see Figure 2.9.
Errors due to the finite width of the autocorrelation
function are illustrated in Figure 2.10; which is a
graphical representation of the Weiner-Hopf equation.
The time interval Δt must be chosen by considering the
highest frequency likely to be present in the SUT's
impulse response. From a theoretical study (Lamb 1970)
into oscillating modes, it has been shown that in order
to detect a peak to within 1%, the ratio of model period
to clock period must be about 20:1. As the ratio
decreases the accuracy falls off rapidly, with a ratio of
10:1 appearing to be a reasonable compromise (accuracy to
within 2.5%).

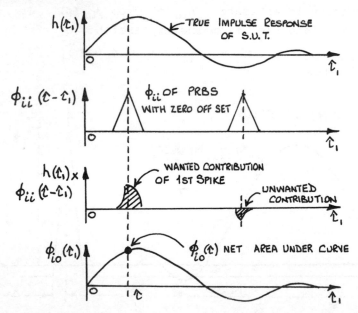

FIG.2.10. Representation of the Weiner-Hopf Equation

Let us consider an example. Suppose we wish to identify
the SUT of Figure 2.11, using PRBS and cross-correlation
to estimate the impulse response. The procedure is as
follows:-

(i)   The observed decay time of the
        complete impulse response is 12
        seconds. Therefore N Δt>12

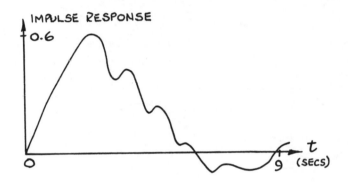

FIG.2.11. Illustrative example - Impulse Response of SUT

(ii)   From the accuracy considerations in
        identifying the secondary mode
        Δt<2 π/(5x10)
        (If we chose 10 clock intervals per
         period of the secondary mode)

If Δt is made equal to 0.05 secs then N should be >240,
so that R = 8 giving N = 255, (2^8 -1) would be
satisfactory for accurate identification if the
instrument is sufficiently flexible to permit such a
choice.In fact for the tests illustrated in Figure 2.12,

the best practical compromise was found to be R = 7 and
Δt = 0.1 secs.   Figure 2.12   also shows the effect of
varying the R and Δt parameters.

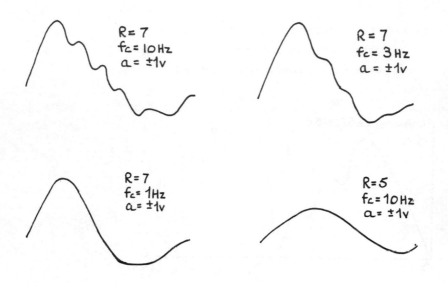

FIG.2.12. Effect on Identification of SUT (Figure 2.11.)
Using Various Values of Δt and R.

2.4.3.   Effect of Measurement Noise and
          Non-linearities

If the corrupting measurement noise n(t) is white, then

$$\phi_{nn}(t) = \rho\delta(t)$$ 2.18

It can then be shown (Hughes,Noton 1962) that the
standard deviation of the uncertainty or error is given
by

$$\sigma = a\sqrt{\{\rho/T\}}$$
2.19

Where T is the total integration time. It is seen that
the error is dependent (although of course non-
deterministically) directly with the signal level, a, and
inversely with the square root of the measurement time.
It is interesting to note that this uncertainty is
independent of the delay time.

If the measurement was undertaken in the presence of
sinusoidal noise, then this can be filtered, all other
characteristics being matched, by selecting the clock
frequency equal to the sinusoidal frequency, as
illustrated in Figure 2.12.

Non-linearities exhibit their presence in a rather
curious and confusing manner, since they are found to
show up on the cross-correlation functions as an apparent
increase in the noise level, the magnitude of which is
not decreased by increasing the integrating time, as
would be the case for broadband noise. In this situation
there are three possible approaches that can be taken:

(i)   The definition of a satisfactory dynamic
        performance can be empirically
        based on the measured cross-correlation
        function, not on the linearised impulse
        response (Towill, Lamb 1970).

(ii)   The effect of the non-linearities can
be reduced by using the previously
mentioned inverse repeat sequences
(Godfrey 1969).

(iii)   The effect of the non-linearities can
be reduced by adding a precisely defined and
synchronised dither signal to the PRBS test
signal (Banasiewitz,Williamson,Lovering 1973).

2.5.   Tolerancing the Impulse Response Estimates
2.5.1.   Introduction

In check-out, we are concerned with devising tests which
classify the SUT as "healthy", i.e. fit for the task for
which it is intended, or "sick", and hence in need of
repair or discarded.

Check-out gates are set relative to the notionalised
'nominal' or 'average' h(t) as measured on the test rig,
thus including all interface effects. Such gates can be
updated based upon operational trials (Brown, Towill,
Payne 1972), and initially set knowing the expected
tolerance for a particular system parameter.

2.5.2.   Setting the Gate Width

Assuming parameter independence (Shooman 1968), the gate
width $\pm g$ at the ith time delay is given by

$$\pm g_i = \left\{ \sum_{j=1}^{J} \left\{ \frac{\delta h(t_i)}{\delta \alpha_j / \alpha_j} \right\}^2 \left\{ \frac{\Delta \alpha_j}{\alpha_j} \right\}^2 \right\}^{\frac{1}{2}} \qquad\qquad 2.20$$

there being J parameters (the α_j) in all affecting h(t).
The partial derivatives are w.r.t. the normalised
parameters, and are the sensitivity functions of h(t),
w.r.t. the parameters. Sometimes the impulse response
sensitivity functions may be calculated directly from the
sensitivity transfer function $\partial H(s)/\partial \alpha_j /\alpha_j$ followed by a
digital simulation. If an adequate mathematical model
does not exist, the impulse response sensitivity
functions can be estimated from breadboard SUT
experiments in which the experiments in which the
parameters are available for adjustments $\partial h(t_i)/\partial \alpha_j /\alpha_j$ are
then estimated via difference methods. The resulting
evaluation of the g_i then results in a 'spread' around
the nominal response, h(t).

2.5.3. Using Frequency Domain Information
2.5.3.1. Introduction

Production test procedures are often based upon frequency
response data but a knowledge of dynamic errors such as
peak impulse error is equally important in assessing the
goodness of a system. Time and frequency domain
responses are related through the Fourier Transform, so
that in designing systems it is not difficult to propose
empirical relationships between performance criteria such
as impulse peak error and bandwidth (Biernson 1956).

2.5.3.2. General Method of Estimating Time Domain
Parameters from Frequency Domain Measurements

Suppose that the (m+n+1) coefficients a_k (k=1,2...n),
b_j (j=1,2...m) of the system transfer function

$$H(s) = \frac{\sum\limits_{j=o}^{m} b_j s^j}{1 + \sum\limits_{k=1}^{n} a_k s^k}$$ 2.21

can vary independently.

Then if

$$\{x_f\} \triangleq \{x_{f_1} \; x_{f_2} \; x_{f_3} \; \ldots\ldots\ldots x_{f_{m+n+1}}\}^T$$ 2.22

is a vector whose elements are an appropriate arbitrary set of measurable frequency domain parameters, such as gain and phase data at specific frequencies,

let

$$\{\frac{\delta x_f}{\delta A}\} \triangleq \begin{bmatrix} \dfrac{\delta x_{f_1}}{\delta a_1} & \dfrac{\delta x_{f_1}}{\delta a_2} & \ldots\ldots & \dfrac{\delta x_{f_1}}{\delta a_n} & \dfrac{\delta x_{f_1}}{\delta b_o} & \ldots & \dfrac{\delta x_{f_1}}{\delta b_m} \\[4mm] \dfrac{\delta x_{f_2}}{\delta a_1} & \dfrac{\delta x_{f_2}}{\delta a_2} & \ldots\ldots\ldots\ldots\ldots\ldots & & & \\[4mm] \vdots & & & & & \\[2mm] \vdots & & & & & \\[4mm] \dfrac{\delta x_{f_{m+n+1}}}{\delta a_1} & \ldots\ldots\ldots\ldots\ldots\ldots\ldots & & & & \dfrac{\delta x_{f_{m+n+1}}}{\delta b_m} \end{bmatrix}$$ 2.23

be the $(m+n+1)$ x $(m+n+1)$ matrix of frequency parameter sensitivity functions. Small changes in frequency domain parameters caused by small changes in system parameters can be defined as a vector:

$$\{\Delta x_f\} \triangleq \{\Delta x_{f_1} \; \Delta x_{f_2} \; \ldots\ldots\Delta x_{f_{m+n+1}}\}^T$$ 2.24

whilst system parameter changes are defined by

$$\{\Delta A\} \triangleq \{\Delta a_1 \quad \Delta a_2 \quad \ldots\ldots \Delta a_n \quad \Delta b_o \ldots\ldots \Delta b_m\}^T \qquad \text{2.25}$$

Hence for first order changes using Equations (2.23,2.24 and 2.25)

$$\{\Delta x_f\} = \{\frac{\delta x_f}{\delta A}\}\{\Delta A\} \qquad \text{2.26}$$

Similarly, let

$$\{\Delta x_t\} \triangleq \{\Delta x_{t_1} \quad \Delta x_{t_2} \quad \ldots\ldots\ldots \Delta x_{t_p}\}^T \qquad \text{2.27}$$

be the (pxl) vector of changes in the time domain vector $[x_t]$

$$\{x_t\} \triangleq \{x_{t_1} \quad x_{t_2} \ldots\ldots\ldots\ldots x_{t_p}\}^T \qquad \text{2.28}$$

which is required to be estimated.
Now

$$\{\frac{\delta x_t}{\delta A}\} \triangleq \begin{bmatrix} \dfrac{\delta x_{t_1}}{\delta a_1} \cdots\cdots\cdots\cdots \dfrac{\delta x_{t_1}}{\delta b_m} \\ \vdots \\ \dfrac{\delta x_{t_p}}{\delta a_1} \qquad\qquad \dfrac{\delta x_{t_p}}{\delta b_m} \end{bmatrix} \qquad \text{2.29}$$

is the px(m+n+1) matrix of time domain parameter sensitivity functions, evaluated at their mean values.

Then using Equations (2.25, 2.27, and 2.29) we can write

$$\{\Delta x_t\} \;=\; \frac{\{\delta x_t\}}{\{\delta A_t\}} \{\Delta A\} \qquad\qquad 2.30$$

Finally using Equations (2.26 and 2.30) we have

$$\{\Delta x_t\} \;=\; \frac{\{\delta x_t\}\{\delta x_f\}}{\{\delta A_t\}\{\delta A_f\}}\{\Delta x_f\} \qquad\qquad 2.31$$

Equation (2.31) gives the fundamental relationship
(Brown, Towill, Payne 1972) between first order changes in
time domain parameters from measured variations in
appropriate frequency domain parameters chosen so that
$\frac{\delta x_f}{\delta A}$ is non-singular.

In a similar manner (Ractliffe 1967) one can develop an
expression linking the variances in the two domains:

$$\{\sigma^2_{x_t}\} \;=\; \{(\tfrac{\delta x_t}{\delta A})^2\}\{(\tfrac{\delta x_f}{\delta A})^2\}^{-1}\{\sigma^2_{x_f}\} \qquad\qquad 2.32$$

where

$$\{\sigma^2_{x_f}\} \;\triangleq\; \{\sigma^2_{x_{f_1}} \;\; \sigma^2_{x_{f_2}} \;\;\ldots\ldots\;\; \sigma^2_{x_{f_{m+n+1}}}\}^T \qquad\qquad 2.33$$

and

$$\{\sigma^2_{x_t}\} \;\triangleq\; \{\sigma^2_{x_{t_1}} \;\; \sigma^2_{x_{t_2}} \;\;\ldots\ldots\;\; \sigma^2_{x_{t_p}}\}^T \qquad\qquad 2.34$$

REFERENCES

Hughes M.T. G. and Noton, A.R.M.
"Measurement of Control System
Characteristics by Means of a
Cross-Correslator"
Proc. Inst Elec. Eng. 109, Part B,
No. 43, p.p. 77-83, 1962

Lamb J.D.   "Use of Pseudo-Random Binary
Sequences for the Production
Testing of Dynamic Systems"
Ph.D. Thesis, U.W.I.S.T., 1970

Towill D.R. and Lamb J.D.
"Pseudo- Random Signals Test
Non-Linear Controls"
Control Engineering, 17, No.11,
p.p. 59-63, 1970

Godfrey K.R.   "The Theory of the Correlation
Method of Dynamic Analysis and
its Application to Industrial
Processes and Nuclear Power Plant"
Measurement and Control, Vol.2,
p.p. 65-70, May, 1969

Banasiewtz H., Williamson S.E. and Lovering, W.F.
"Use of Synchronised Dither in
Pseudo-Random-Sequence Testing"
Electronics Letters, Vol.19,
p.p. 334-5, July 1973

Brown J.M., Towill D.R. and Payne P.A.
                "Predicting Servomechanisms
                Dynamic Errors"
                Radio and Electronic Engineer,
                Vol,42, No.1, p.p. 7-20, 1972

Shooman M.L.   "Probabilistic Reliability: An
                Engineering Approach"
                McGraw-Hill Book Co., New York,
                Chapter 7 1968

Biernson G.A.   "A General Technique for
                Approximating Transient Response
                from Frequency Response
                Asymptotes"
                Trans Amer. Inst. Elec. Eng.
                (Applications & Industry), 75,
                Part 2, p.p. 253-73, 1956

Ractliffe J.F. "Elements of Mathematical
                Statistics"
                Oxford University Press, 1967

CHAPTER 3
Basic Ideas
of Fault Location

3.1. Introduction

We now develop some basic ideas of selecting the test
features and means of representing the response
deviations. In general, using the transfer function
representation of the system implies that the analysis in
both time and frequency domains is of a continuous
nature. Whether a system passes the go/no go check out
test depends on the relationship between the system's
deviation from nominal and the check out gates. So we
will be processing the deviation response in either
domain - see Figure 3.1.

44

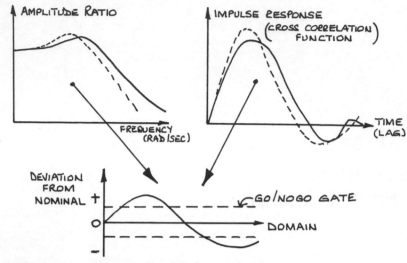

FIG 3.1.   System deviation response

Using a digital computer, perhaps as an integral part of
an ATE station, means that we have to sample the
deviation response. This in turn results in having to
make the decision on which should be the test features.
However, we will first assume that we have selected the
features and consider methods of establishing the
recognition matrix.

3.2.   Quantitisation

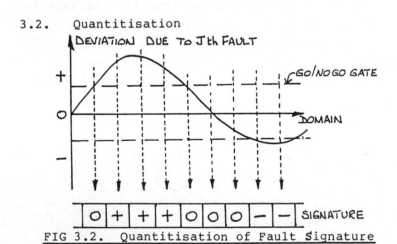

FIG 3.2.   Quantitisation of Fault Signature

The simplest way of establishing a signature is to simply
establish whether the deviation is above, below or within
the go/no go gates. Thus in Figure 3.2 we see that at
the selected features a pattern of 0,+,-, can be used to
represent the Jth fault. Alternatively,0,+1,-1 can be
used leading to a diagnostic algorithm based upon
arithmetic operations rather than logical operations.

To establish the signature corresponding to a particular
fault, which is then grouped with the other signatures to
form the recognition matrix, the transfer function must
be adjusted to reflect the fault being simulated. In the
simulation,what constitutes the "fault" must be precisely
defined, although not necessarily precisely in fault
level. The easiest faults to consider in general are
those that result in a single unambiguous fault state,
i.e. the component value is of a binary nature - good or
bad. If this condition exists then the fault signature
is unique for that fault.However, if a component value in
its failed state can take any value over and above a
given level then the fault signature is not unique to the
fault but depends on the fault level; see Figure 3.3.

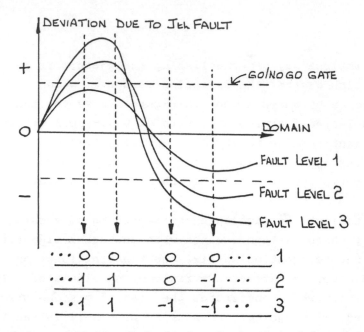

Fig 3.3.   Effect of fault level on the signature

It is evident that to cover fully the fault shown
diagramatically in Figure 3.3 at least 3 signatures would
be required to allow for the range in possible deviations
shown. Conceptually, this is possible, particularly as
memory costs decrease. However, when one considers the
other fault cases it is easy to see how confusion and
lack of coherence could result.

The key to whether a robust scheme is possible is the
examination of the recognition matrix. The two main
criteria are:

  (i)   entries for individual faults are
          unique;
  (ii)   there is no confusion as the fault
          levels change.

In many systems it is impossible to distinguish between
particular faults using only the response deviation
vector. This is due to the inherent characteristics of
the design of the system. In other words the cause and
effect relationships between the components and response
are identical, and so indistinguishable under fault
conditions.

3.3.   Recognition Matrix

$$R = \begin{bmatrix} x_{11} & x_{12} & x_{13} & \cdots\cdots\cdots x_{1N} \\ x_{21} & x_{22} & x_{23} & \cdots\cdots\cdots x_{2N} \\ x_{31} & x_{32} & x_{33} & \cdots\cdots\cdots x_{3N} \\ \cdot & \cdot & \cdot & \cdot \\ \cdot & \cdot & \cdot & \cdot \\ x_{M1} & x_{M2} & x_{M3} & \cdots\cdots\cdots x_{MN} \end{bmatrix} = \begin{bmatrix} x_{iJ} \end{bmatrix}$$

$$\quad\quad\quad 3.1$$

$$i=1,2\ldots M$$
$$J=1,2\ldots N$$

The recognition matrix is made up of response vectors xJ
for each likely fault:

$$R = \begin{bmatrix} x_1 \vdots x_2 \vdots x_3 \vdots \cdots \vdots x_N \end{bmatrix}^T \quad\quad\quad 3.2$$

In general the number of likely faults is established
first (N) and then the least number of test points (M)
established to enable adequate fault diagnostics to be
implemented.
At first sight a diagonal recognition matrix would seem
the optimum; giving a one to one correspondence between a
test point and a particular fault. This would result in
R being not only diagonal but square, i.e. N faults by N
measurements. This condition would certainly be
unambiguous. However, it would not be an efficient use
of the test measurements.

48

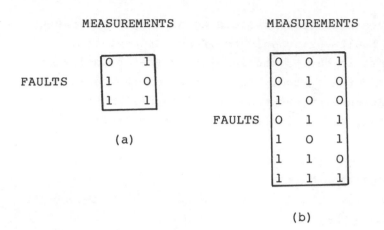

FIG 3.4   Efficient Use of Test Data Patterns

Figure 3.4   gives examples of quantitised entries for two
and three measurement points.   Even ignoring negative
entries, the possible information content is better than
a one to one correspondence between measurements and
faults.

The situation indicated in Figure 3.4 is ideal.   In
practice not all the possible patterns would be
realizable but a better than one to one relationship
should be aimed for.

3.4.   Selection of Test Features

The natural approach in feature selection is to choose
that time delay or frequency that produces the greatest
deviation, from nominal, in the system's response for a
given fault.   If we initially assume that positive and
negative changes in a parameter produce symmetric
deviations then this approach will result in one feature

per likely fault. Whilst using the maximum deviation   as
a selection criterion intuitively is correct, it does not
guarantee discrimination between faults. However, it
does afford a starting point in feature selection. The
subsequent recognition matrix can be examined and
features adjusted to give the required discrimination.
In the frequency domain, a starting point is to use the
asymptotic log plot versus frequency, choosing
frequencies between break points. If the system has
resonances then further frequencies are chosen at the
break frequencies,(Garzia 1971).

3.5.   Estimation of Likely Fault

The simplest approach in estimating the likely fault is
template matching. This compares element by element of
the SUT's deviation vector with the stored vectors making
up the recognition matrix. The stored vector giving the
highest number of matching elements indicates the likely
fault. Whilst the approach is attractive from the
computational point of view (using 1's and 0's is
arithmetically trivial), it is not robust under
conditions of varying fault levels and variations due to
the production tolerances of the non-faulty components.
Of course, one does not require the deviations to be
quantitised to use a template matching approach. Such a
method is the cross-product which will be considered
later. Generally speaking, for quantitised deviations
one requires a probabilistic approach. One such method
using quantitised data is that due to Sriyananda, who
used a combination of voting and probabilistic
information to rank likely faults. This method will be
considered in detail in the next chapter.

50

3.6.   The Cross-Product Fault Location Method
3.6.1. Introduction

This method is the simplest, conceptually, to
implement. To generalise we will assume that the data
has not been quantitised, enabling us to introduce the
concept of normalisation.

3.6.2. The Method

Let (Xij) be the recognition matrix i=1...M,j=1...N of a
system with N likely faults having M tests performed.

Let Yi, i=1...M be the deviation vector of the SUT. Then
we can form N cross products Fj j=1...N
where

$$F_j = \sum_{i=1}^{M} x_{ij} y_i \qquad\qquad 3.3$$

Then the Fj with the largest magnitude indicates the most
likely fault and the next largest, the next most likely
fault, and so on over the range of likely faults.
Computationally, this method poses little difficulty.
Intuitively, the approach agrees with the concept of
statistical correlation.

3.6.3. An example

Consider Table 3.1 which represents the recognition
matrix of a system tested using PRBS with the associated
cross-correlation function.

TIME DELAY (ms)

F			0	5	10	15	20	25	30	35	40
A	J1	(xi1)	0	-1.5	0	0.5	0.5	0.4	0.3	0.2	0.1
U	J2	(xi2)	0	-3.0	-3.0	-1.5	0.5	1.5	1.6	1.4	1.0
L	J3	(xi3)	0	0.6	1.5	2.4	3.0	2.5	2.0	1.5	1.0
T											
S											

TABLE 3.1. Recognition Matrix for SUT

For a particular SUT the following deviation vector was obtained:

(yi)   0 -1.0 -0.5 1.0   0   0.9 -0.2 0.7 -0.4

Then the resulting cross products are

F1 = 2.40   F2 = 4.61   F3 = 3.55

Hence the most likely fault is J2.

3.6.4. Normalisation

Whilst there will be a high correlation between the test deviation vector and the corresponding recognition matrix vector, there will also be high values in the F's due simply to the magnitudes of the vectors.
Figure 3.5 illustrates the problem. Whilst it seems highly likely that the fault is J2, the size of the deviation of J3 will swamp the discrimination,making it unlikely that J2 will ever be diagnosed. In fact we are really interested in relative shapes or patterns not simply magnitudes.

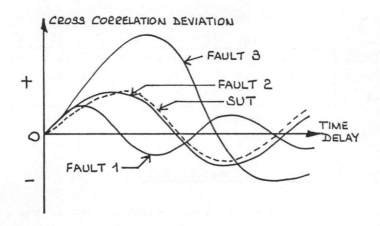

FIG 3.5. Deviation vector and recognition vectors of SUT

A solution is to normalise the cross products, viz.

$$F_j^* = \frac{\sum\limits_{i=1}^{M} x_{ij} y_i}{||x_{ij}||_E} \qquad \qquad 3.4$$

where $||xij||_E$ is the Euclidean norm of $[xj]$

$$||x_{ij}||_E = [x_{1j}^2 + x_{2j}^2 + \ldots\ldots + x_{Mj}^2]^{\frac{1}{2}} \qquad 3.5$$

Returning to the example in 3.6.3 we find

F1 = 1.37   F2 = 0.87   F3 = 0.64

Now we see that the most likely fault is J1.

CHAPTER 4
Quantitised Methods of Fault Location

4.1.   Introduction

In this chapter we consider the method due to Sriyananda
(Sriyananda, Towill 1972).   It is classed as a voting
technique.   It uses pattern recognition type properties
of the SUT, but uses this information to assign "votes"
on likely failure causes, thus combining the better
aspects of these two basic techniques.

4.2.   Derivation of the Basis of the Diagnostic Scheme

Consider a system, whose response R is a function of N
variables v (which correspond to the values of the
components of the system, or to a set of parameters that
define the system):

$$R = R(d,v)$$
$$v = \begin{bmatrix} v_1 & v_2 & v_3 & \cdots\cdots & v_N \end{bmatrix}^T$$

$$d = \text{domain, time or frequency}$$

4.1

If $v \to v_f$ under fault conditions, then the faulty response is given by

$$R_f = R(d, v_f) \qquad\qquad 4.2$$

$$\text{Let} \quad y(d) = \Delta R = R_f - R \qquad\qquad 4.3$$

and $x(d) = v.\nabla R$ where ∇ is evaluated w.r.t. v

$$\text{i.e.} \quad x_j(d) = \frac{\delta R(d,v)}{\delta v_j / v_j} \qquad\qquad 4.4$$

Now define

$$z = \phi_{xy}(0)$$

where

$$z_j = \phi_{x_j y}(0) = \int_\infty^\infty x_j(d) y(d) dd \qquad\qquad 4.5$$

Assuming the x's are uncorrelated

i.e.

$$\phi_{x_i x_j}(0) = 0 \quad \text{for} \quad i \neq j \qquad\qquad 4.6$$

then, for small changes

$$z_j = \begin{cases} > 0 & \text{for } v_{fj} > v_j \\ = 0 & \text{for } v_{fj} = v_j \\ < 0 & \text{for } v_{fj} < v_j \end{cases}$$

By natural laws, the zero condition of Equation 4.6 is never realisable. However, it may be assumed that the element j which corresponds to the largest numerical value of z_j is more likely to be faulty than any other, given that all elements were equally likely to be faulty before the test results were known.

4.3. Discretisation

In the physical test situation, continuous test domains
are unlikely. Assuming M test points, then the
discretised form of Equation 4.5 would be

$$z = [x]^T y \qquad\qquad 4.7$$

i.e. $\quad z_j = \sum\limits_{i=1}^{M} x_{ij} y_i$

and the assumption of 4.6. becomes

$$\sum_{i=1}^{M} x_{ij} x_{ik} = 0, \text{ for } j \neq k \qquad\qquad 4.8$$

It would generally be sufficient to choose the features
so as to make

$$\sum_{i=1}^{M} (x_{ij})^2 > \max \left| \left\{ \sum_{i=1}^{M} x_{ij} x_{ik} , k \neq j \right\} \right| \qquad 4.9$$
$$j=1,2...N$$

If this is not possible, it would mean more access points
are needed to be able to differentiate between those
parameters for which it is not true.

To simplify the procedure both the x and y are
quantitised

Let

$$x_{ij}^* = \begin{cases} 1 & \text{for } x_{ij} > E_o \\ 0 & \text{for } |x_{ij}| < E_o \\ -1 & \text{for } x_{ij} < -E_o \end{cases} \qquad 4.10$$

and

$$y_i^* = \begin{cases} 1 & \text{for } y_i > E_1 \\ 0 & \text{for } |y_i| < E_1 \\ -1 & \text{for } y_i < -E_1 \end{cases} \qquad 4.11$$

Where Eo and El are predetermined fault levels

Now a modification to the definition of z given by (4.7) is introduced. If x*ij = 0 and y*i = 0 the contribution to z evaluated according to (4.7) is zero. However, this combination of values contains the information that the variable j is unlikely to be faulty. (This argument cannot be extended to the case where x*ij = 0 and y*j = 0 due to the errors involved in the assumption of 4.8.) Hence, define

$$
z_{Fj}^{*} = \sum_{i=1}^{M} x_{ij}^{*} y_{i}^{*}
$$

$$
z_{Nj}^{*} = \sum_{i=1}^{M} |x_{ij}^{*}| (1-|y_{i}^{*}|)
$$

$$
z_{j}^{*} = \begin{cases} (\text{sgn}.z_{Fj}^{*}) \{|z_{Fj}^{*}|-z_{Nj}^{*}\} & \text{for } |z_{Fj}^{*}| > z_{Nj}^{*} \\ 0 & \text{for } |z_{Fj}^{*}| < z_{Nj}^{*} \end{cases}
$$

$$4.12$$

Then z* defined by (4.12) using (4.10) and (4.11) is used to establish the fault location.

4.4. Fault Probability Ranking

If P is the 'a priori' relative probability that the parameters are faulty, then the 'a posteriori' relative probability P* may be obtained from the above as

$$
P_{j}^{*} = \begin{cases} \dfrac{P_{j} z_{j}^{*}}{\sum_{i=1}^{N} P_{i}|z_{i}^{*}|} & \text{for } \sum_{i=1}^{N} P_{i}|z_{i}^{*}| > 0 \\ 0 & \text{for } \sum_{i=1}^{N} P_{i}|z_{i}^{*}| = 0 \end{cases}
$$

$$j=1,2...N \qquad 4.13$$

Thus, if a fault condition is detected, the faults are
ranked as to indicate the relative probability of each
fault; and if no fault is detected, all the fault
probabilities are set to zero.

4.5.   A Pattern Recognition Interpretation

The approach can be interpreted as a pattern recognition
technique (Fu 1968).

$y*$ is the feature vector, spanning the feature space Ay,
say.   Then aj (j=1,2...N) are the N possible pattern
classes corresponding to a fault in component j, thus
constituting the fault space.   Considering single fault
cases, these constitute N mutually exclusive regions.
The discriminant function associated with the decision
process can be seen to be $P*$ as defined in 4.13.

4.6.   Feature Selection

It has been shown (Toussaint 1971) that the test sub-set
of features need not contain the best single feature.
The Sriyananda method obtains an "optimum" sub-set of
features by truncating an "ordered" set of features.
This ordering indicates the degree of differentiation
that a particular feature would be capable of.

Sriyananda proposes a "goodness criterion" for a feature
as

$$g_i \quad = \quad \log_N \{\max\{P(a_j/x^*_{ij}), \; j=1,2....N\}\} + 1 \qquad \textbf{4.14}$$

Being a logarithmic form it has additive properties.

The criterion gi as defined will be zero if P(aj/x*ij) is equal to l/N which is the "worst case", corresponding to equal probability of failure for all components. The criterion will be unity if P(aj/x*ij) is equal to one; being the "best case", corresponding to a capability of detecting at least one particular fault with almost certainty.

Using Bayes Theorem, (4.14) can be re-written as

$$g_i = \log_N \{\max\{\frac{P(x^*_{ij}/a_j)\ P(a^*_j)}{P(x^*_{ij})} , \quad j=1,2...N\}\} + 1$$

$$\simeq \log_N \{\max\{\frac{P(x^*_{ij}/a_j)\ P(a_j)}{P(x^*_{ij})} , \quad j=1,2...N\}\} + 1 \qquad 4.15$$

As x*ij can only take values of O,+1,-1, then this can be interpreted as

$$g_i = \log_N \{\max\{\frac{|x^*_{ij}\ P(a_j)|}{\sum\limits_{j=1}^{N}|x^*_{ij}|P(a_j)} , \quad j=1,2...N\}\} + 1$$

$$= \log_N \{\frac{\max(P_j,j=1,2...N)\ \text{for non zero}\ x^*_{ij}}{\sum\limits_{j=1}^{N}|x^*_{ij}|P_j}\} + 1 \qquad 4.16$$

4.7.   Voting Logic

Of the quantitised methods, combining as it does a pattern recognition approach with probabilistic criteria, it probably affords one of the best techniques for fault location. It easily lends itself to implementation on a computer. Table 4.1. gives a summary of the voting element and Figure 4.1. gives the flow chart for computer implementation of the voting logic.

	Above Tolerance Band yi=+1	Within Tolerance Band yi=0	Below Tolerance Band yi=-1
High Positive xij=+1	Vote for J High	Vote for J near Nominal	Vote for J Low
Relative Insensitive xij=0	No Votes Cast	No Votes Cast	No Votes Cast
High Negative xij=-1	Vote for J Low	Vote for J near Nominal	Vote for J High

Table 4.1. Sriyananda's Voting Logic

Implied in this method is that a component or parameter
is faulty with the sign of accumulated votes indicating
whether the component's value is above or below nominal.
Thus in establishing the recognition matrix an average of
the positive and negative deviations is used. This is an
efficient use of test data and is a reasonable approach
if the positive and negative deviations are symmetric
about the nominal response. However, if this symmetry is
absent, then the efficiency of the diagnosis is naturally
reduced. If the positive and negative changes in
component value are treated as two distinct cases, then
the size of the recognition matrix is immediately
doubled. Thus the two approaches can be simplistically
presented as either average the positive and negative
changes and use many test points or treat them separately

60

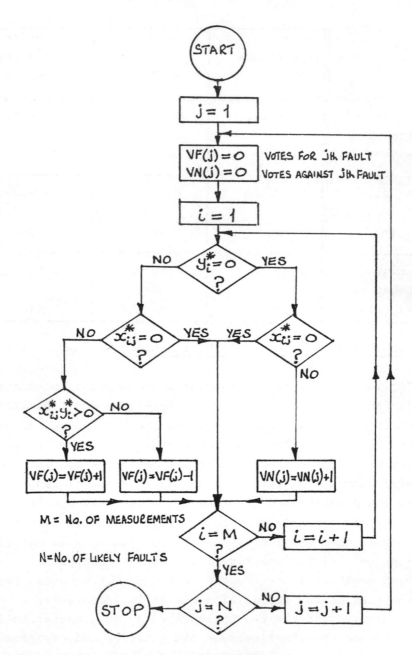

FIG. 4.1. Sriyananda's Voting Logic

and reduce the number of test points to produce a
manageable recognition matrix. We will return to this
problem later.

4.8.   Training or Learning

If the structure and the component values of the healthy
system are completely known, then the $x*j(j=1,2...N)$
could be determined using (4.4) and (4.10.). In the
absence of such complete information, a process of
"learning" is a more suitable method for generating the
$x*j$. This entails the study of a sample of cases, either
actual or simulated, where both the response and fault
are known.   Even in the situation where the healthy
system is completely known, this approach is useful when
the likely fault level is relatively large; for the
assumption in (4.4) is that the changes are small (in the
calculus sense).
Assuming that the sample is representative, (4.10) and
(4.11) can be simplified by setting Eo=El=E . It is
then possible to generate a recognition matrix [x*] for
any value of E.   This adds another variable to the
diagnostic scheme, wherein the effect of various values
of E can be determined so the diagnostic level can be
optimised.

4.9.   An Example
4.9.1. The System

The system considered is an existing complex
electro-hydraulic control system. An analogue computer
simulation together with some of the actual physical
components was used, see Figure 4.2.

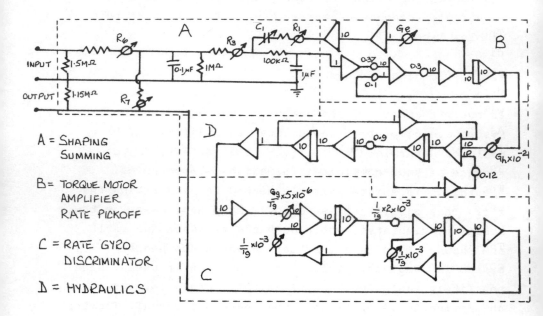

FIG. 4.2. Analogue Simulation of Electro-Hydraulic Servo

The nominal, high and low values of the five components
and four parameters chosen as the most likely to fail
under fault conditions are shown in Table 4.2.

Fault	μF Cl	kΩ Rl	kΩ R3	kΩ R6	kΩ R7	Ge	Gh	Gg	secs Tg
Nominal	8.0	6.8	270	150	150	0.7	50	50	1/400
High	12.0	7.5	300	165	165	0.77	55	55	1/360
Low	4.7	6.1	230	135	135	0.63	45	45	1/440

Table 4.2 Nominal & Fault Values of Likely Faults of SUT

4.9.2. Test Domain

The impulse response was used as the basis of the
diagnosis (the frequency response, step response or any
other characteristic of the system could be used if so
desired).The response was obtained by cross-correlation,
with the input being PRBS (as described in Chapter 2).
As the system is known to have a significant high
frequency response around 50Hz.,a clock frequency of
300Hz was used in the generation of the input sequence.
The peak signal level was   ±2V., and 10 shift registers
were used, giving a sequence length of 1023. Each
observation was obtained by convolution over 10 cycles.
The resolution of the observation was 0.10V.,
corresponding to 0.15 for the impulse response.

The simulation was found to be very noisy, as is the real
system, and 20 runs of the nominal system response and 5
runs each of the faulty responses were recorded so as to
facilitate an acceptable learning operation. Figure 4.3
shows two examples of the responses obtained.

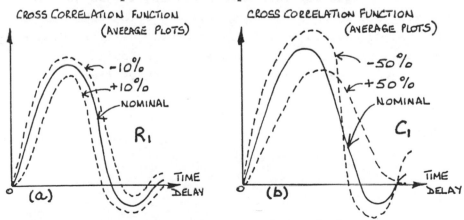

FIG. 4.3. Cross Correlation Functions of the SUT

The difficulty of correct diagnosis with this type of
fault is evident from Figure 4.3a.; not only are the
differences in the response of the system with the
different faults insignificant in most cases, but they
are also of the same order of magnitude as the
differences between repeated runs of the nominal system.
The only exceptions were the response faults due to the
value of the capacitance, as shown in Figure 4.3b.

4.9.3. Learning and Fault Location

The data collected during the simulation was first used
to design the diagnostic scheme, by the use of learning
techniques discussed in section 4.8. They were then
treated as data from "unknown" systems, i.e. the systems
could be sick or healthy.
For brevity, it is assumed that all faults have equal
probabilities of occurrence (1/9). The value of E was
varied to observe its effect on the diagnosability of the
scheme. Table 4.4 shows the recognition matrix for E=1.
Once the recognition matrix was derived, it was used to
identify the imposed faults in turn. As the fault level
was changed the diagnosability naturally varied. Figure
4.4 shows the relationship between diagnosability and
fault level, keeping the number of features fixed at 31.

I	TIME	C	R1	R3	R6	R7	Ge	Gh	Tg	Gg	G(I)
1	0.0	0	0	0	0	0	0	0	0	0	0.000
2	10.0	0	0	0	0	0	0	0	-1	0	1.000
3	20.0	0	1	-1	-1	1	-1	1	0	1	0.114
4	30.0	-1	1	-1	-1	1	-1	1	1	1	0.000
5	40.0	-1	1	0	-1	1	-1	1	0	0	0.185
6	50.0	-1	0	-1	-1	1	0	0	0	1	0.268
7	60.0	-1	0	1	-1	1	1	0	0	-1	0.185
8	70.0	-1	-1	1	-1	1	1	-1	0	0	0.114
9	80.0	0	-1	1	0	1	1	-1	0	-1	0.185
10	90.0	1	-1	1	-1	0	1	-1	0	-1	0.114
11	100.0	1	-1	1	0	0	1	-1	0	-1	0.185
12	110.0	1	-1	1	0	0	1	0	0	-1	0.268
13	120.0	1	0	1	0	0	1	-1	0	-1	0.268
14	130.0	1	0	1	0	0	1	0	0	0	0.500
15	140.0	1	0	0	0	0	0	-1	0	0	0.685
16	150.0	1	0	0	0	0	0	0	0	0	1.000
17	160.0	1	0	0	0	0	0	0	0	0	1.000
18	170.0	1	0	0	0	0	-1	0	0	0	0.685
19	180.0	0	1	-1	0	0	-1	0	0	0	0.500
20	190.0	0	1	-1	0	0	-1	1	0	1	0.268
21	200.0	0	1	-1	0	0	-1	0	0	1	0.369
22	210.0	0	1	-1	0	0	-1	0	0	0	0.500
23	220.0	0	0	-1	0	0	-1	0	0	1	0.500
24	230.0	-1	0	-1	0	0	-1	1	0	0	0.369
25	240.0	-1	0	-1	0	0	0	0	0	0	0.685
26	250.0	-1	0	-1	0	0	0	1	0	0	0.500
27	260.0	-1	0	0	0	0	0	0	0	0	1.000
28	270.0	-1	0	0	0	0	0	0	0	0	1.000
29	280.0	-1	0	0	0	0	0	0	0	0	1.000
30	290.0	0	0	0	0	0	0	0	0	0	0.000
31	300.0	0	0	0	0	0	0	0	0	0	0.000

TABLE 4.4 Recognition Matrix for case E=1

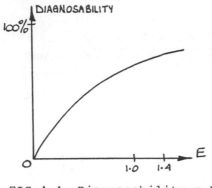

FIG.4.4. Diagnosability v E

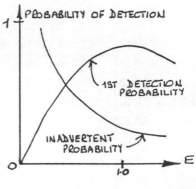

FIG.4.5. Detection v E

The results of a Monte Carlo simulation are shown
diagrammatically in Figure 4.5. Results shown are
collective, in that success or failure in diagnosis is
considered over all the simulated faults. In general,
incorrect diagnosis falls off with increasing fault
level, whilst for this simulation, detection using the
highest vote reaches a maximum, then falls off.

4.10   Generalisation of the Voting Technique
        (Sriyananda, Towill, Williams 1975)
4.10.1 Introduction

The basic voting technique can be interpreted as a
pattern recognition technique. The simplest form of
pattern recognition is template matching (Seshu, Waxman
1966) (Maenpea, Stelham, Stahl 1069) (Sriyananda,
Towill, Williams 1973) in which the range of possible
measurements at each test is graded into a number of
divisions, each coded into a logical order. Each fault
case is then represented by a signature composed of the
coded row vector. Measurements on the SUT are similarly
graded and compared term by term with each stored vector
until agreement is found.

The objections to the template approach are that
measurement noise and drift in the non-faulty components
quickly reduce the diagnosability to a low level. In
addition it only attempts to find a unique match which is
often unattainable.

4.10.2 Template Voting

To add the ranking characteristics of a voting technique
to the template matching method would eliminate the need
for an exact match. The modification is trivial, merely
voting for the fault if the vector elements matched and
against if they did not.
Using the quantitised levels defined by 4.10 and 4.11,
the fault diagnosis index can be written for

$$z^{*}_{Fj} = \sum_{i} x^{*}_{ij} y^{*}_{i} + \sum_{i} (1 - |x^{*}_{ij}|)(1 - |y^{*}_{i}|)$$

$$z^{*}_{Nj} = M - z^{*}_{Fj} \qquad \text{where M= no. of features}$$

as

$$z^{*}_{j} = 2z^{*}_{Fj} - M \quad \text{if } (2z^{*}_{Fj}-M) > 0$$

$$\phantom{z^{*}_{j}} = 0, \quad \text{otherwise}$$

4.17

4.10.3 Feature Selection

As an alternative to Sriyananda's method of selection of
the test features one could choose tests that have the
maximum deviations. This will generally lead to M ~ N.
Sriyananda also considers both positive and negative
changes in a component value as a single fault, using the
average value of the quantitised deviations. During
diagnosis, the sign of the cumulative vote indicates the

direction of deviation. With template voting, these two
cases are considered as two distinct faults, which
results in twice the number of fault cases, because the
deviations caused by positive and negative changes are
generally not symmetrical.

4.10.4. Illustrative Example

Figure 4.6 shows a 7 component passive network used as a
test vehicle.

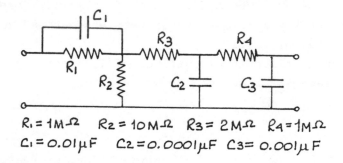

$R_1 = 1M\Omega \quad R_2 = 10M\Omega \quad R_3 = 2M\Omega \quad R_4 = 1M\Omega$
$C_1 = 0.01\mu F \quad C_2 = 0.0001\mu F \quad C_3 = 0.001\mu F$

FIG. 4.6. Circuit Used in Simulation Study

The circuit is typical of those considered by Garzia
(Garzia 1971). Table 4.5 shows the recognition matrix
obtained using Sriyananda's method (for convenience
measurements are listed downwards). For this recognition
matrix Equation 4.8 has been evaluated, see Table 4.6.
The calculations have been normalised by dividing each
correlation coefficient $\phi(xj,xk)$ by $\sum_i (xij)^2$. By
inspection it can be seen that in 38 cases out of 42,
$\sum_i (xij)^2 \geqslant |\phi(xi,xj)|$ suggesting that the recognition
matrix is based on a reasonable set of test frequencies.
The four combinations where $\sum_i (xij)^2 = |\phi(xi,xj)|$ are
C1R3, C1C2, C1C3 and R1R2. In each case there are a

Freq.(r/sec)	Measurement	R1	R2	R3	R4	C1	C2	C3	gi%
10	Amp.Ratio	-1	1	0	0	0	0	0	64
20	Amp.Ratio	-1	1	0	0	0	0	0	64
30	Amp.Ratio	-1	0	0	0	0	0	0	100
40	Amp.Ratio	-1	0	0	0	0	0	0	100
50	Amp.Ratio	-1	0	0	0	0	0	0	100
50	Phase	0	0	-1	0	0	0	-1	64
60	Phase	0	0	-1	0	0	0	-1	64
70	Phase	0	0	-1	0	0	0	-1	64
80	Amp.Ratio	0	0	-1	0	0	-1	-1	44
80	Phase	0	0	-1	0	0	0	-1	64
90	Amp.Ratio	0	0	-1	0	0	-1	-1	44
90	Phase	0	0	-1	0	0	0	-1	64
100	Amp.Ratio	0	0	-1	0	0	-1	-1	44
100	Phase	0	0	-1	0	0	-1	-1	44
200	Amp.Ratio	0	0	-1	0	1	-1	-1	29
200	Phase	0	0	-1	0	0	-1	-1	44
300	Amp.Ratio	0	0	-1	0	0	-1	-1	44
2000	Amp.Ratio	0	0	-1	0	0	0	0	100
2000	Phase	0	0	0	-1	0	-1	-1	44
3000	Phase	0	0	0	-1	0	-1	-1	44
4000	Phase	0	0	0	-1	0	-1	0	64
5000	Phase	0	0	0	-1	0	-1	0	64
6000	Phase	0	0	0	-1	0	0	0	100
7000	Phase	0	0	0	-1	0	0	0	100
8000	Phase	0	0	0	-1	0	0	0	100

TABLE 4.5. Recognition Matrix for Sriyananda's Voting
Method for System shown in fig.4.6

substantial number of distinguishing data points despite the apparently high correlation.

	R1	R2	R3	R4	C1	C2	C3
R1	100	100	0	0	0	0	0
R2	-40	100	0	0	0	0	0
R3	0	0	100	0	-100	-64	86
R4	0	0	0	100	0	36	14
C1	0	0	-8	0	100	-9	-7
C2	0	0	54	57	-100	100	64
C3	0	0	92	29	-100	82	100
$\sum_i x^2{}_{ij}$	5	2	13	7	1	11	14
$\max\lvert \phi(x_j, x_k)\rvert$	2	2	12	4	1	9	12

TABLE 4.6. %Correlation between Fault Cases for
Frequency Data Points of Table 4.5.

For the template voting 7 frequencies were chosen;the corresponding recognition matrix is shown in Table 4.7.

Freq. r/s	Meas't	R1 +	R1 -	R2 +	R2 -	R3 +	R3 -	R4 +	R4 -	C1 +	C1 -	C2 +	C2 -	C3 +	C3 -
20	A.R.	-1	1	1	-2	0	0	0	0	0	0	0	0	0	0
40	A.R.	-1	2	0	-2	0	0	0	0	0	0	0	0	0	0
50	Phase	0	0	0	0	-1	1	0	0	0	0	0	0	-1	1
200	A.R.	0	0	0	0	-2	2	0	0	1	-2	-2	2	-2	2
400	A.R.	0	0	0	0	-2	2	-1	1	0	-1	-2	2	-2	2
2000	Phase	0	0	0	0	0	2	-2	2	0	0	-1	2	-1	2
7000	Phase	0	0	0	0	0	0	-1	2	0	0	0	2	0	1

TABLE 4.7. Recognition Matrix for Template Matching

Up to 5 quantitised states were used. For both
recognition matrices the fault level was set at +,- 50%.
Three simulations were carried out, one implementing
Sriyananda's scheme, one implementing the template voting
and for completeness, one using template matching. For
each case four fault levels were considered, +,-35%,
+,-50%, +,-70% and random changes drawn from a normal
distribution with mean equal to the nominal component
value and std. dev. equal to 30% of nominal. Each fault
(14 in total) was simulated 100 times, thus giving a
total of 16,800 simulations. In each simulation the non-
faulty component values were drawn from a normal
distribution having a mean equal to the nominal value and
std. dev. equal to 3% of the nominal.

	+,-70%			+,-50%			+,-35%			Random		
	S	TV	TM	S	TV	TM	S	TV	TM	S	TV	TM
R1+	100	45	8	100	100	95	92	98	86	91	90	78
R1-	100	100	0	100	100	66	100	100	0	92	87	7
R2+	16	98	17	62	99	66	31	61	42	33	88	40
R2-	0	100	4	0	100	88	1	1	0	16	40	22
R3+	91	56	33	97	80	67	86	94	0	43	66	19
R3-	0	73	0	0	69	34	54	100	0	23	52	4
R4+	100	88	38	100	97	25	100	98	0	82	54	11
R4-	100	88	38	100	97	25	100	98	0	82	54	9
C1+	88	100	73	99	100	80	100	100	76	94	98	68
C1-	81	2	0	100	78	62	100	37	9	98	38	16
C2+	4	41	23	94	99	76	69	88	26	46	70	30
C2-	1	13	13	94	98	97	69	52	0	54	33	15
C3+	91	97	14	78	55	45	27	0	0	24	18	11
C3-	100	100	0	99	100	83	3	19	0	19	21	7

       869,1013,246   1122,1269,913   920,872,239   780,800,337

S=Sriyananda,TV=Template Voting,TM=Template Matching

TABLE 4.8. % Successful Diagnosis under Simulation

The results clearly show the superiority of the voting
schemes over template matching. There is no significant
difference between the voting schemes. Whilst the
Sriyananda scheme uses many features, 25, it uses only 3
quantitised states and 7 faults, while the template
voting used 5 quantitised states and 14 faults and 7
features.
In the next chapter non-quantitised methods will be
considered.

REFERENCES

H. Sriyananda & D.R. Towill
"Fault Diagnosis via Automatic
Dynamic Testing - A Voting
Technique"
The Automation of Testing, IEE
Conf. Publication No.91, p.p. 196-
201, Sept 20-22 1972

K.S. Fu   "Sequential Methods in Pattern
Recognition and Machine Learning"
Academic press, New York, 1968
Chapter 1

G.T. Tousaint "Note on the Selection of
Independent Binary Valued Features
for Pattern Recognition"
IEE Trans on Information Theory,
Vol.IT-17, Sept 1971, p.618

H. Sriyananda, D.R. Towill, J.H. Williams
"Voting Techniques for Fault
Diagnosis from Frequency-Domain
Test Data"
IEE Trans Rel Vol.R-24, p.260-267,
Oct 1975 [Reprinted in IEEE Press
Selected Reprint Series "The World
of Large Scale Systems", edited by
J.D. Palmer, R. Saeks 1982]

S. Seshu, R. Waxman
"Fault Isolation in Conventional
Linear Systems - A Feasibility
Study"
IEEE Trans Rel, Vol.R-15, p.11-16,
May 1966

J.H. Maenpaa, C.J. Stehman, W.J. Stahl
"Fault Isolation in Conventional
Linear Systems: A Progress Report"
IEEE Trans Rel Vol.R-18, p.12-14,
Feb 1969

H. Sriyananda, D.R. Towill, J.H. Williams
"Selection of Test frequencies for
System Fault Diagnosis"
AIAA Paper No.70-864, AIIA
Guidance and Control Conference
Key Biscayne, Fl. USA, Aug 1973

R.F. Garzia   "Fault Isolation Computer Methods"
NASA Contractor Report, CR 1758
Feb 1971

CHAPTER 5

Non-Quantitised Methods
of Fault Location

5.1.   Introduction

We now consider in more detail methods that use test data
in a non-quantitised form. Whilst quantitised data is in
a convenient form for computerisation for both storage
and processing, the very act of quantitising removes
information. This loss in information requires, in
general, replacement in the form of using many features.
In fact, Sriyananda suggests three times as many features
as likely faults. So if we argue for retention of the
actual response deviation as a value, methods have to be
developed to efficiently use such data that can cater for
faults of varying levels.

76

5.2. Linearised Key Element Search Method
(L. Buchshaum et al, 1964)

5.2.1. Theoretical Development

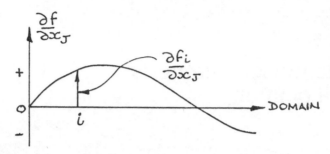

FIG. 5.1. Typical Sensitivity Function of
Measurement w.r.t. Component

Assume a deviation vector yi [i=1,2...M] for a SUT. If we
have the rate of change of response w.r.t. each likely
fault, $\partial fi/\partial xj$ [j=1,2...N]. Then if Δx_J is the change
from nominal of the Jth component, we can form a
performance index.

$$S_J = \sum_{i=1}^{M} (y_i - \frac{\delta f_i}{\delta x_J}.\Delta x_J)^2 \qquad\qquad 5.1$$

where we wish to find Δx_J that will make S_J a minimum.

Hence

$$S_J = \sum_{i=1}^{M} (y_i^2 - 2\frac{\delta f_i}{\delta x_J} y_i \Delta x_J + (\frac{\delta f_i}{\delta x_J})^2 (\Delta x_J)^2) \qquad\qquad 5.2$$

Therefore

$$\frac{dS_J}{d\delta x_J} = -2 \sum_{i=1}^{M} \frac{\delta f_i}{\delta x_J} y_i + 2\sum_{i=1}^{M} (\frac{\delta f_i}{\delta x_J})^2 \Delta x_J \qquad\qquad 5.3$$

For a minimum, $\dfrac{dS_J}{d\Delta x_J} = 0$ therefore

$$\Delta x_J = \frac{\sum\limits_{i=1}^{M} \frac{\delta f_i}{\delta x_J} y_i}{\sum\limits_{i=1}^{M} (\frac{\delta f_i}{\delta x_J})^2} \qquad 5.4$$

[$\dfrac{d^2 S_J}{d\Delta x_J^2} = 2 \sum\limits_{i=1}^{M} (\frac{\delta f_i}{\delta x_J})^2$ which is clearly positive, and so the turning point is a minimum]

Substituting the expression for Δx_J into (5.2) gives

$$S_J = \sum_{i=1}^{M} y_i^2 - 2\sum_{i=1}^{M} (\frac{\delta f_i}{\Delta x_J})y_i \left[\frac{\sum\limits_{i=1}^{M}\frac{\delta f_i}{\delta x_J} y_i}{\sum\limits_{i=1}^{M}(\frac{\delta f_i}{\delta x_J})^2}\right] + \sum_{i=1}^{M}(\frac{\delta f_i}{\delta x_J})^2 \left[\frac{\sum\limits_{i=1}^{M}\frac{\delta f_i}{\delta x_J}}{\sum\limits_{i=1}^{M}(\frac{\delta f_i}{\delta x_J})^2}\right]^2$$

$$S_J = \sum_{i=1}^{M} y_i^2 - \frac{\left[\sum\limits_{i=1}^{M}(\frac{\delta f_i}{\delta x_J}) y_i\right]^2}{\sum\limits_{i=1}^{M}(\frac{\delta f_i}{\delta x_J})^2} \qquad 5.5$$

Now $\sum\limits_{i=1}^{M} y_i^2$ is not dependent on J, so that for a minimum value of S_J we require to maximise K_J where

$$K_J = \frac{\left|\sum\limits_{i=1}^{M}(\frac{\delta f_i}{\delta x_J}) y_i\right|^2}{\sum\limits_{i=1}^{M}(\frac{\delta f_i}{\delta x_J})^2} \qquad 5.6$$

Then K_J can be used to rank the likely faults, with the most likely being given by the maximum K_J , the next most likely being associated with the next highest of K_J and so on.

Figure 5.2 gives a schematic of the procedure.

78

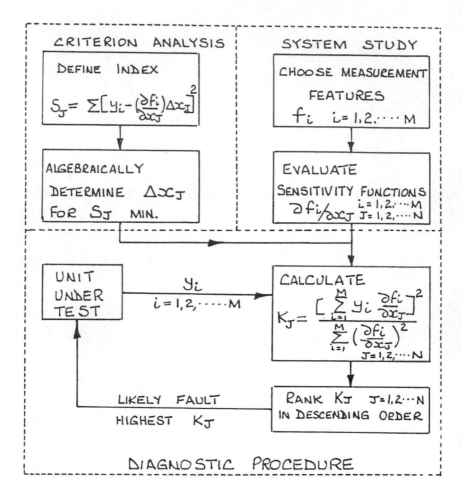

FIG 5.2. Key Element Search Data Flow Schematic

5.2.2. Example

Table 5.1 shows the recognition matrix based upon the
sensitivity functions for three parameters of a system at
five selected time delays of the system's cross
correlation function.

	Measurements fi				
Time Delay (sec)	0.05	0.07	0.14	0.18	0.20
fault x1	0.01	0.05	-0.02	0.06	-0.01
fault x2	0.06	0.08	0.02	-0.02	-0.01
fault x3	0.01	0.05	0.01	0.13	0.05

Table 5.1. Sensitivity Values $\partial fi/\partial x_j$

The SUT gave the following deviations from nominal.

Time Delay (sec)	0.05	0.07	0.14	0.18	0.20
Deviations	0.20	0.24	0.18	-0.16	-0.08

Thus $K1 = (0.0196)/(0.166) = 0.0023$

$K2 = (0.0300)/(0.0109) = 0.138$

$K3 = (0.0072)/(0.032) = 0.0016$

So the faults are ranked in the order:- K2;K1;K3.

5.3. Nearest Neighbour Rule (Towill, Williams 1977)
5.3.1. Introduction

The key element search method uses values of the sensitivity functions at selected test points to form the recognition matrix. Instead of using the calculus of small variations the recognition matrix can be established using deviations due to chosen changes in the system's parameter values. The recognition matrix can now be considered as being made up of a set of vectors representing the fault conditions. Subsequently,given the deviation vector of the SUT, the problem resolves itself into one of matching the SUT deviation vector with each vector of the recognition matrix in turn. A convenient method is to use a metric or distance measure between vectors, choosing the 'nearest' vector as the indication of the likely fault.

5.3.2. The Method

Let $x_J = [x_{J1}, x_{J2}, \ldots x_{JM}]^T$ represent the deviation
vector due a fault in parameter J, and $y = [y_1, y_2, \ldots y_M]^T$
represent the deviation vector of the SUT. Then one can
form N distances d_J defined as

$$d_J^2 = \sum_{i=1}^{M} (x_{Ji} - y_i)^2$$

5.7

Then the likely fault is assumed to be d_J^2 min. In fact
the d_J^2 can be ranked in descending order giving the more
practical solution of indicating the order in which to
investigate the likely fault.

5.3.3. Normalisation

The method presented so far relies on the deviations of
the SUT to be of a similar magnitude to the corresponding
stored fault vector.

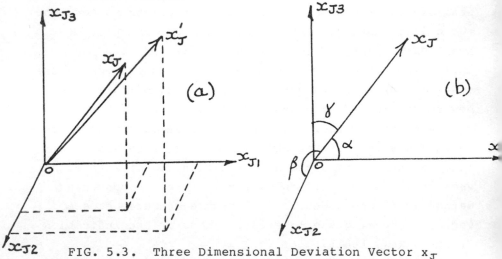

FIG. 5.3. Three Dimensional Deviation Vector x_J

A three dimensional example as shown in Figure 5.3(a) x_J represents one of the vectors of the recognition matrix. As the magnitude of the change in the Jth parameter changes so the x_J will move to x'_J, say. This results in a poor level of diagnosability when the fault level of the SUT differs considerably from the fault level used in setting up the recognition matrix. If we consider Figure 5.3(b) which indicates the angular measures of x_J with respect to the three axes, then

$$\alpha = \cos^{-1}\left[\frac{x_{J1}}{\sqrt{(x_{J1}^2 + x_{J2}^2 + x_{J3}^2)}}\right]$$ 5.8

with similar expressions for the angles β and γ

$$\cos\alpha = x_{J1}/|x_J|$$
$$\cos\beta = x_{J2}/|x_J|$$
$$\cos\gamma = x_{J3}/|x_J|$$ 5.9

These are the direction cosines of the vector x_J. Now as x_J changes to x'_J then the changes in these direction cosines will only depend on the relative changes in x_{Ji} and not on their absolute values. Thus using direction cosines will result in a recognition matrix that represents, in a normalised form, a wide range in parameter changes. Of course, the deviation vector y of the SUT, must also be normalised in the same manner. Thus the normalised scheme becomes

$$\bar{x}_{Ji} = x_{Ji}/|x_J| \qquad i=1,2..M; \ j=1,2..N \qquad 5.10$$

and

$$\bar{y}_i = y_i/|y| \qquad i=1,2..M \qquad 5.11$$

The d_J^2 can then be ranked in descending order suggesting the likely probability of the faults.

5.4.   Data processing Optimisation

5.4.1. Introduction

This is a convenient time to consider the amount of data
processing that is involved in implementing the
techniques we have considered. The scale of operations
is directly proportional to the size of the recognition
matrix. One dimension is fixed - the number of fault
cases considered. The other dimension depends upon the
number of features or test points used. Feature selection
is considered in Chapter 6.

The quantitised methods have the advantage of having many
zeros in the recognition matrix.This enables sparse
matrix techniques to be used with the consequent
reduction in computational time. Non-quantitised
methods, such as the nearest neighbour technique,at first
sight do not have these characteristics. However, each
fault vector in the recognition matrix will be made up of
significant and non-significant elements.The significant
elements will be those corresponding to the test
measurements exhibiting high sensitivity for the
particular fault.
We now consider various approaches taking into account
these points.

5.4.2. Pareto Analysis

Consider a fault vector $x_J = [x_{J1}, x_{J2}, \ldots x_{JM}]^T$, then the
$|x_{Ji}|$ can be ranked in descending order and a cumulative
sum formed. Figure 5.4 shows the arrangement. A
significance level can be set, as shown, subdividing the
elements into significant and non-significant elements.

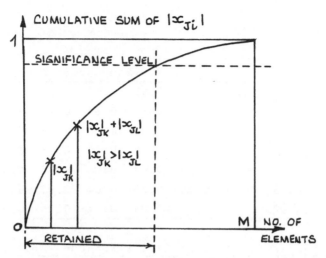

FIG. 5.4.   Pareto Analysis of the Elements of x_J ,
            a Fault Vector

The non-significant elements are then set to zero. This
is repeated for the N fault cases, giving a sparse
recognition matrix. The problem is what level to set the
significance threshold. The usual approach is to adjust
this level as the diagnosis proceeds, paying off
diagnosability against the reduction in non-zero entries
in the recognition matrix.

Example

Table 5.2 shows a recognition matrix for a system with
seven likely faults, and eight features have been chosen.
The fault vectors have been normalised so their elements
sum to unity. Figure 5.5 shows the Pareto analysis for
the seven vectors. It is seen from the plots that taking
>0.8 as the cut-off point will result in a number of
insignificant elements. It is not necessary to use the
same cut-off level for each fault vector, but in this
case it is a reasonable approach.

Faults	Features	(Measurements)					
x1	0.30	0.25	0.25	0.05	0.04	0.03	0.04	0.04
x2	0.07	0.07	0.16	0.14	0.14	0.15	0.21	0.06
x3	0.04	0.06	0.05	0.05	0.20	0.25	0.20	0.15
x4	0.07	0.06	0.14	0.23	0.13	0.15	0.15	0.07
x5	0.24	0.34	0.22	0.04	0.04	0.05	0.04	0.03
x6	0.25	0.20	0.35	0.05	0.04	0.05	0.03	0.03
x7	0.07	0.03	0.06	0.04	0.14	0.16	0.23	0.27

Table 5.2. Recognition Matrix for the Example

Retaining the significant elements and setting to zero
the insignificant elements leads to the recognition
matrix as shown in Table 5.3.

Faults	Features	(Measurements)					
x1	0.03	0.25	0.25	0	0	0	0	0
x2	0	0	0.16	0.14	0.14	0.15	0.21	0
x3	0	0	0	0	0.20	0.25	0.20	0.15
x4	0	0	0.14	0.23	0.13	0.15	0.15	0
x5	0.24	0.34	0.22	0	0	0	0	0
x6	0.25	0.20	0.35	0	0	0	0	0
x7	0	0	0	0	0.14	0.16	0.23	0.27

Table 5.3. Reduced Recognition Matrix

85

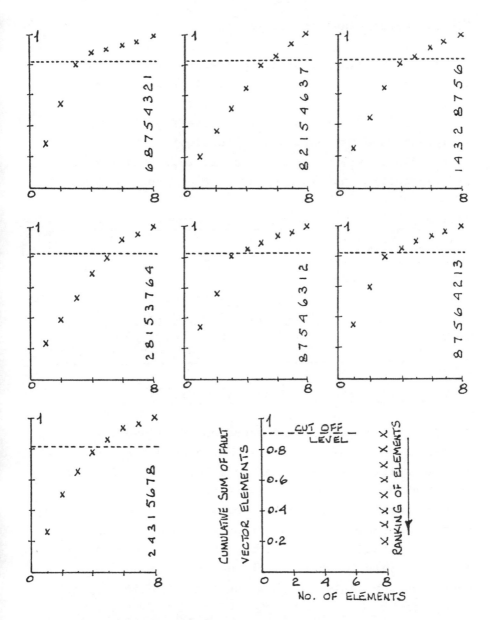

FIG.5.5. Pareto Analysis of Fault Vectors of
Recognition Matrix shown in Table 5.2.

5.4.3. Coarse/Fine Techniques

Of course, one is not bound by the arbitrary order of the
likely faults. If Table 5.3 is examined it is seen that
there are basic patterns of non-significant and
significant entries. By grouping these patterns a more
efficient data processing is possible.
If the faults are re-ordered to: x1,x5,x6,x2,x4,x3,x7
the recognition matrix becomes that   shown in Table 5.4

Faults   Features   (Measurements)

Fault								
x1	0.03	0.25	0.25	0	0	0	0	0
x5	0.24	0.34	0.22	0	0	0	0	0
x6	0.25	0.20	0.35	0	0	0	0	0
x2	0	0	0.16	0.14	0.14	0.15	0.21	0
x4	0	0	0.14	0.23	0.13	0.15	0.15	0
x3	0	0	0	0	0.20	0.25	0.20	0.15
x7	0	0	0	0	0.14	0.16	0.23	0.27

(Blocks labelled A, B, C)

Table 5.4. Regrouped Recognition Matrix

This enables a coarse/fine approach to be used. The
deviation vector of the SUT is normalised and the
non-significant elements set to zero. Initially, the
actual values of the retained elements are not critical,
rather where are they in the vector. This means matching
to a sub set of the vectors represented by the three
blocks outlined in Table 5.4. After the corresponding
block has been identified, the nearest neighbour
technique can be used within the block to identify the
likely fault.

5.4.4. Sequential Testing

The regrouped recognition matrix also affords the
possibility of implementing the tests or features in a
sequential manner. For instance, considering Table 5.4,
if there was a significant deviation at the first test
point then this would indicate that the fault was
associated with block A, and tests 4 to 8 were not
required. For the example given, tests 1, 4 and 8 would
give indications of whether the fault lay in areas
covered by blocks A, B or C respectively.

One problem in taking this approach is that to get to the
regrouped matrix requires the complete fault vectors to
be formed and the Pareto analysis undertaken. This
becomes less critical if the range of fault levels is not
wide, so that the raw deviations are enough to indicate
significance.
Also, if the test domain is either a frequency response
or a cross correlation measurement then it is often not
practical to step singly through the test points.
Sequential testing is more likely to be successful where
there is a "mixed" test schedule, that includes
mechanical and physical checks as well as dynamic and
static measurements.

REFERENCES

L. Buchshaum, M. Dunning, T.J.B. Hannon, L. Moth
             "Investigation of Fault Diagnosis
             by Computational Methods"
             Pennington, Rand AD-601204, 1964

D.R. Towill and J.H. Williams
             "Fault Diagnosis of a Complex
             Electro-hydraulic System"
             Proc of Int. Conf. on Technical
             Diagnosis, Prague 1977

CHAPTER 6
Optimisation of Feature Selection

6.1. Introduction

One of the most important aspects of any fault location method is the selection of features or measurements. Two fundamental criteria have to be considered. One is the requirement to use as few tests as possible, the other is to achieve reasonable fault cover. With proper and efficient feature selection, diagnosability can be increased considerably. Improper or ineffective features will increase the number of features necessary for testing,thereby increasing test time and/or computational facilities required and also increase confusion and errors in diagnosability. Feature selection is the optimum retention of a minimum number of measurement points or features while maintaining and/or maximising the probability of correct diagnosis (Andrews 1972).

One natural approach is to use those measurements for which the response deviations are a maximum for a given fault. However, the best set of features may not necessarily contain the best individual feature (Toussant 1971). We will now consider a method of feature selection which considers the features obtained from the complete dynamic range of the system and selects an optimum sub set (Varghese, Towill, Williams 1978).

6.2. Discriminating Power of Features

Suppose Sij represents the ith measurement of the jth class or fault and there are M such measurements and N classes. Then for the ith feature the discriminating information in the feature space (all possible features) is given by:

$$D_i^* = \bigcup_{j=1}^{N} \overline{S_{ij} \cap S_{i(j+1)}} \cup \overline{S_{ij} \cap S_{i(j+2)}} \cup \cdots \cdots \overline{S_{ij} \cap S_{iN}} \qquad 6.1$$

This information is now estimated from the information contained in the recognition matrix. A distance or metric is used

$$D = \left[\sum_{j=1}^{N-1} (x_{ij} - x_{i(j+1)})^2 \right]^{\frac{1}{2}} \qquad 6.2$$
$$i=1,2\ldots M$$

This is a measure of the effectiveness of each feature to discriminate between fault cases and is used in the optimisation technique as one of the criteria for the retention of features.

6.3. Separability Measure of Fault Cases

The separation between fault cases can be represented as:

$$S(k,l) = S_{ik} \cap S_{il} \qquad\qquad 6.3$$

A set of features has a separability measure for each
combination of fault cases and the ability of the tests
to discriminate between two cases depends upon the
distance between the cases. This distance is formulated
for each combination and will reflect the fault case
separability of the tests.

Then simply this measure is:

$$d(k,l) = \left[\sum_{i=1}^{M} (x_{ik} - x_{il})^2 \right]^{\frac{1}{2}} \qquad\qquad 6.4$$

where $0 \leqslant d(k,l) \leqslant 2$ (as the xijs' are normalised)
Thus there are $N(N-1)/2$ such distances for N faults.
Figure 6.1 indicates diagrammatically a typical
calculation of $d(k,l)$.

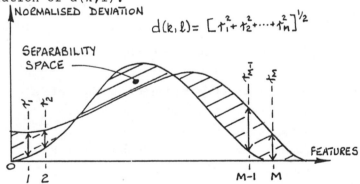

FIG. 6.1. Separability Space and Measure of two Faults.

This is a measure of discrimination between fault cases
and if zero, the chosen feature set is unable to
distinguish between the two fault cases, k and l. The
combination of fault cases can be ranked according to its
separability measure.

6.4. Confidence Level of a Feature Set

The most accurate way of measuring the merit of a set of
features is to determine the diagnosability using the
feature set and to compare this with that obtained using
other sets. This can be accomplished using a simulation
via a breadboard model or a dynamic mathematical model.
It is seldom convenient to evaluate all possible subsets
of features in this manner as it is computationally
prohibitive. It is therefore advantageous to utilize a
quality factor or confidence level capable of evaluating
features at a level where computational requirements are
less demanding i.e. at the feature selection stage rather
than at the diagnostic stage.

The suggested method is to calculate the confidence level
of a set of tests specifically considering the following
questions:-

    i)   What constitutes an effective set of
           tests?

    ii)   How is this effectiveness dependent on
           the correlations among the various
           combinations of fault cases?

    iii) How will this be as a measure of the
           diagnosability?

A set of tests have a separability measure for each
combination of fault cases considered two at a time, and
the ability of the tests to discriminate between two
cases depends upon the "distance" between them. There is
thus an obvious similarity between the separability
measure and the nearest neighbour rule of fault location.

A formula is adopted which is capable of indicating the
non-discriminatory components without being dominated by
combinations of components of high separability measure.
A confidence level of unity is assigned to combinations
of fault cases with separability measures of 50% or more.
Those combinations with separability measures less than
50% retain their actual values. If C1 and C2 are the
confidence levels thus obtained, total percentage
confidence level is defined as:

$$C = \left[\frac{C_1 + C_2}{N(N-1)/2}\right] \times 100 \quad \text{for N fault cases.} \qquad 6.5$$

Although the percentage of confidence could be increased
by lowering the 50% margin of separability measure, this
limit is chosen purposely as a method of reducing the
effect of variations of non-faulty components. This
enables the confidence level to be an effective measure
of the efficiency of test features even for a noisy
system. A general flow chart for the calculation of
confidence level from separability measures is shown in
Figure 6.2

6.5.   Optiminsation Technique for Feature
       Selection

As the first step of optimisation, the features of very
little information content are discarded. If all elements
in a column, i.e. the deviations of response for all fault
cases from nominal for one feature of the normalised
recognition matrix, are within a preset tolerance, that
column (feature) is discarded. This preset tolerance is
proposed to be 10% of $\sum_{j}^{M}|xij|$, though it can be varied.

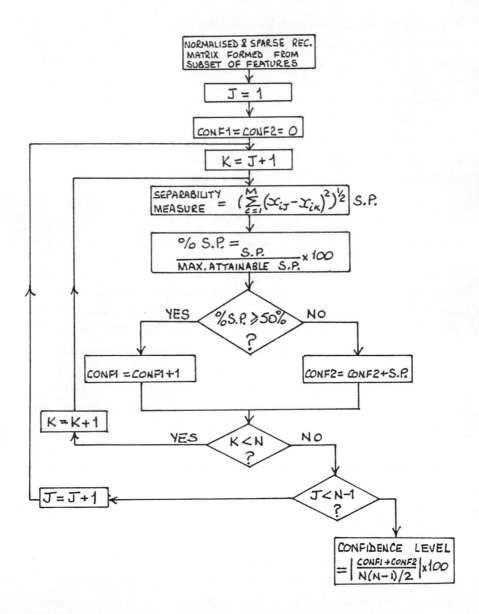

FIG.6.2. Flow Chart for Separability Measure and
Confidence Level.

The features retained can be called the "first optimum set". A new recognition matrix is set up for this feature set which is normalised and made sparse. Optimisation of this second feature set is carried out as follows:-

Let $\quad F(J) = |x_{iJ} - x_{(i+1)J}| \qquad J= 1,2...N$

Let $\quad SUMC(I) = \sum_{i=1}^{N} |x_{iJ}|$

and $\quad SUM = SUMC(I) + SUMC(I+1)$

Then if F(J), J=1,2...N is less than some fraction, 1/p, say, of the SUM, the retentions of both features need not increase the discriminatory information and hence the feature of less discriminatory power is discarded (Varghese 1977). The factor p is kept a variable to add or discard features to achieve maximum confidence level. The technique is shown in Figure 6.3.
Fault location is then achieved using this optimised recognition matrix and the nearest neighbour rule.

6.6.   Illustrative Example

To illustrate the optimisation techniques the test circuit used in Chapter 4 was utilised. Figure 6.4. shows the test circuit together with its transfer function.
For each value of p (see Figure 6.3) a subset of features is retained. These features are then used to form a recognition matrix which is normalised and made sparse. The separability measures of combinations of fault cases and hence confidence level were calculated.

96

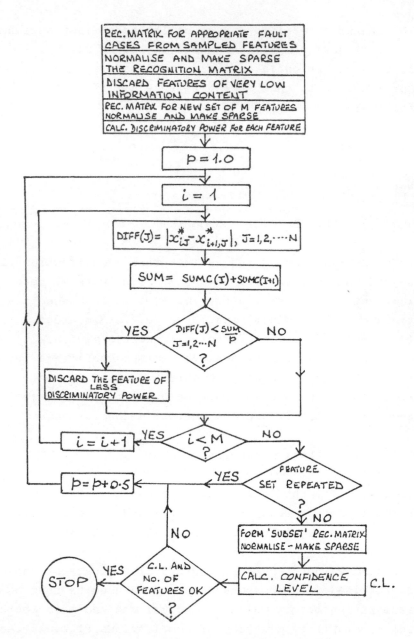

FIG.6.3. Flow Chart for Feature Optimisation.

The optimisation was continued until a satisfactory
confidence level was obtained.

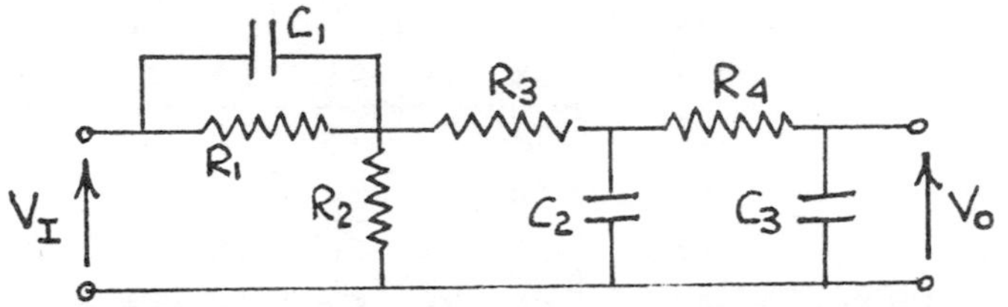

$$R_1 = 1M\Omega \quad R_2 = 10M\Omega \quad R_3 = 2M\Omega \quad R_4 = 1M\Omega$$
$$C_1 = 0.01\mu F \quad C_2 = 0.0001\mu F \quad C_3 = 0.001\mu F$$

$$\frac{V_0(s)}{V_I(s)} = \frac{a_0 + a_1 s}{b_0 + b_1 s + b_2 s^2 + b_3 s^3}$$

$$a_0 = R_2 \quad a_1 = R_1 R_2 C_1 \quad b_0 = R_1 + R_2$$
$$b_1 = R_1 R_2 C_1 + R_2 R_4 C_3 + R_1 R_4 C_3 + (C_2 + C_3)(R_2 R_3 + R_1 R_3 + R_1 R_2)$$
$$b_2 = R_1 R_2 R_3 C_1 (C_2 + C_3) + R_1 R_2 R_4 C_3 (C_1 + C_2) + R_3 R_4 C_2 C_3 (R_1 + R_2)$$
$$b_3 = R_1 R_2 R_3 R_4 C_1 C_2 C_3$$

FIG.6.4. Passive Circuit Used as Test Case.

For the simulation, it was assumed that only one
component was faulty at a time and all faults were
equally likely to occur. The non-faulty components were
allowed to vary within a tolerance limit. The faults
were randomly drawn from a normal distribution of
standard deviation 30% of the nominal values. The
non-faulty components were drawn from normal
distributions having standard deviations of 3%, 6% and 9%
in turn, of the nominal values. Each fault situation was
simulated 100 times. For each subset of features selected
during the optimisation procedure, diagnosability was
calculated for all 9 fault cases considered.Fifteen such
subsets were considered making 40,500 simulations in all.

Measurements rad/sec. A=Amp.Ratio P=Phase	Confidence Level %	Average Diagnosability First Choice when: $\sigma_1 =$			$\sigma_2 =$		
		3%	6%	9%	3%	6%	9%
100(P)	50	21	19	19	20	20	18
10(A),100(P)	68	48	45	38	46	37	29
10(A),100,600(P)	86	86	73	60	79	56	39
10(A),100,600,10000(P)	87	90	75	62	83	60	41
10(A),100,600,9000(P)	88	90	76	66	80	60	51
10(A),100,600,8000(P)	88	90	76	53	80	63	54
10(A),100,600,7000(P)	88	93	77	66	82	52	48
10,80,1000(A), 100,600,6000(P)	88	95	84	65	89	56	47
10,80,500,900(A), 100,600,6000(P)	88	94	82	60	86	69	58
10,80,200,400,700(A), 100,600,5000,9000(P)	89	96	83	69	86	67	47
10,70,200,400,700(A), 100,600,5000,9000(P)	89	96	81	74	87	58	58
10,70,200,400,700(A), 100,600,4000,8000(P)	90	96	82	67	89	55	47
10,70,200,400,700(A), 100,200,600,4000,7000(P)	89	96	83	69	84	55	53
10,70,200,400,600,1000(A), 100,200,600,4000,7000(P)	89	96	81	69	85	53	49
10,60,200,400,600,1000(A) 100,200,600,3000,6000, 9000(P)	89	96	82	69	84	53	51

SEE FIG.6.6 FOR EXPLANATION OF σ_1 , σ_2

Table 6.1. Confidence Levels and Diagnosability of Various Measurement Sets.

As an additional test, the whole procedure was repeated
with the non-faulty component values chosen to be greater
than one standard deviation away from their nominal
values i.e. using the tails of the distributions. Hence
91,000 simulations were carried out. Results are
tabulated in Table 6.1.

Figure 6.5 shows the confidence level against the number
of subsets of features.

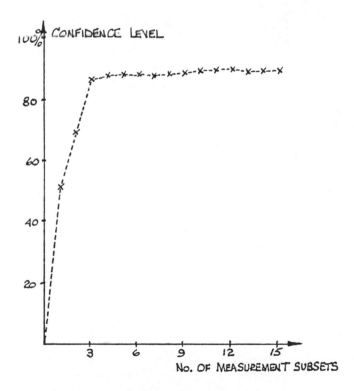

FIG.6.5. Measurement Subsets Selected During Optimisation
Against Confidence Level.

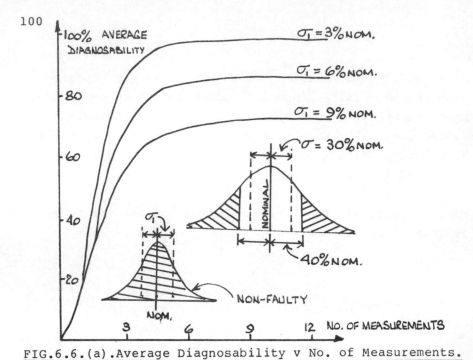

FIG.6.6.(a).Average Diagnosability v No. of Measurements.

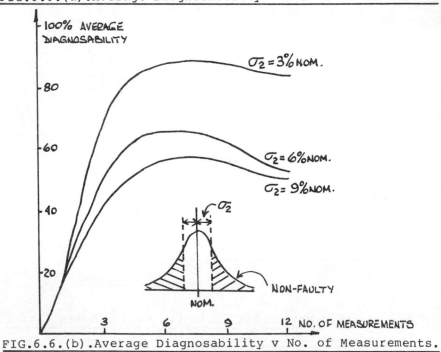

FIG.6.6.(b).Average Diagnosability v No. of Measurements.

Whilst Figure 6.6. shows the effect of the number of
measurements on average diagnosability, confidence level
against average diagnosability is shown in Figure 6.7.,
indicating that the confidence level is a good guide for
feature selection.

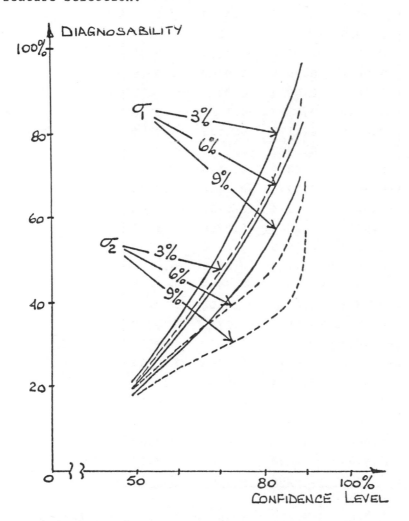

FIG.6.7. Effect of Confidence Level on Diagnosability.

102

A typical recognition matrix is shown in table 6.2.

	Amp.Ratio 10	Phase 100	Phase 600	Phase 9000 (rad/sec.)
R1-	0.0441	-0.0322	-0.0096	0.0000
R1+	-0.0397	0.0333	0.0027	0.0000
R2-	-0.0747	0.0408	0.0153	0.0011
R3+	-0.0013	-0.1353	-0.1246	-0.0197
R4+	-0.0002	-0.0416	-0.1200	-0.0708
C1-	0.0001	-0.0558	0.0465	0.0066
C1+	0.0002	0.0224	-0.0214	-0.0022
C2+	-0.0012	-0.0813	-0.1094	-0.0530
C3+	-0.0016	-0.1124	-0.1123	-0.0354
R1-	0.7952	-0.5809	-0.1741	-0.0001
R1+	-0.7652	0.6417	0.0526	0.0000
R2-	-0.8638	0.4717	0.1768	0.0128
R3+	-0.0071	-0.7314	-0.6736	-0.1065
R4+	-0.0013	-0.2862	-0.8254	-0.4867
C1-	0.0020	-0.7650	0.6376	0.0903
C1+	0.0080	0.7210	-0.6893	-0.0708
C2+	-0.0083	-0.5558	-0.7482	-0.3622
C3+	-0.0098	-0.6904	-0.6900	-0.2172
R1-	0.7952	-0.5809	-0.1741	0.0000
R1+	-0.7652	0.6417	0.0000	0.0000
R2-	-0.8638	0.4717	0.1768	0.0000
R3+	0.0000	-0.7314	-0.6736	0.0000
R4+	0.0000	-0.2862	-0.8254	-0.4867
C1-	0.0000	-0.7650	0.6376	0.0000
C1+	0.0000	0.7210	-0.6893	0.0000
C2+	0.0000	-0.5558	-0.7482	-0.3622
C3+	0.0000	-0.6904	-0.6900	-0.2172

Table 6.2.Recognition Matrices For Feature Set Amp.Ratio
@ 10 r/s ,Phase @ 100,600 and 9000 r/s.

Fall-off in diagnosability as the variability of the
non-faulty components increases is shown in fig.6.8.

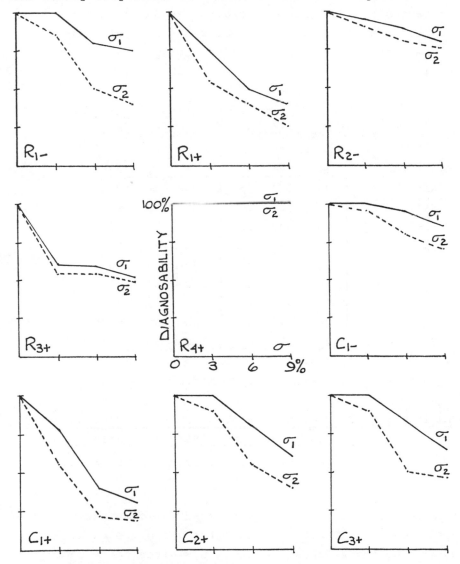

FIG.6.8. Fall in Diagnosability as Variability of
Non-faulty Components Increases.

104

A considerable increase in diagnosability is obtained by
considering the faults of the minimum and next minimum
distances i.e. 1st and 2nd choices as shown in Table 6.3,
since a high probability of locating the fault within
these components may be perfectly acceptable in large
scale systems.

	% Diagnosibility					
	1st.Choice			1st.& 2nd.Choice		
	$\sigma_i =$			$\sigma_i =$		
Fault Case	3%	6%	9%	3%	6%	9%
R1-	99	79	76	99	88	87
R1+	75	53	38	100	91	79
R2-	98	93	82	100	99	88
R3+	63	63	52	100	95	86
R4+	100	100	100	100	100	100
C1-	100	86	85	100	96	85
C1+	78	38	33	85	52	49
C2+	100	81	67	100	100	93
C3+	100	81	65	100	99	98

σ_i Std.Dev. of non-faulty components

Table 6.3. % Diagnosis of Random Faults (σ =30% of nom.)
using the Measurement Set: Amp.Ratio @ 10
rad/sec. and Phase @ 100,600 9000 rad/sec.

The separability measure calculated for combinations of
fault cases in Table 6.4 indicates which are the
confusing fault cases and Table 6.5 gives three cases as
an example. The confusing fault cases are those with low
separability measure.

No.	Fault Case Combination		Separability Measure %
1	R1-	R1+	99.56
2	R1-	R2-	99.79
3	R1-	R3+	47.67
4	R1-	R4+	58.74
5	R1-	C1-	57.67
6	R1-	C1+	80.61
7	R1-	C2+	52.29
8	R1-	C3+	48.93
9	R1+	R2-	13.23
10	R1+	R3+	85.81
11	R1+	R4+	76.94
12	R1+	C1-	86.42
13	R1+	C1+	51.75
14	R1+	C2+	82.34
15	R1+	C3+	84.96
16	R2-	R3+	85.64
17	R2-	R4+	80.03
18	R2-	C1-	79.03
19	R2-	C1+	62.50
20	R2-	C2+	83.50
21	R2-	C3+	85.07
22	R3+	R4+	33.94
23	R3+	C1-	65.91
24	R3+	C1+	72.92
25	R3+	C2+	20.53
26	R3+	C3+	11.11
27	R4+	C1-	80.89
28	R4+	C1+	56.41
29	R4+	C2+	15.34
30	R4+	C3+	25.21
31	C1-	C1+	99.94

Table 6.4. Separability Measure of Fault Combinations

106

Fault Case Combination		Separability Measure	% Faults Confused with σ_l =		
Diagnosed	Confused	%	3%	6%	9%
R1+	R2−	13.22	25	32	39
R1+	R4+	76.94		6	2
R3+	C3+	11.11	37	43	23
C2+	R4+	15.34		5	12
C2+	C3+	10.31		14	14

σ_l Std.Dev. of Non-faulty Components

Table 6.5. % Faults with Confused Diagnosis (1st. choice)

It is seen for this circuit that a subset of 3 features
achieved a reasonable and satisfactory overall
diagnosability above which an increase in the number of
features does not make any considerable increase in
diagnosability. Also, the effect of the number of
features on diagnosability is indexed by the confidence
level prior to the diagnostic stage, thus proving to be
an efficient performance index or figure of merit and the
separability measure of combinations of faults is a
measure of confusion and indicates the confusing fault
cases.

REFERENCES

Andrews H.C.   "Introduction to Mathematical
                Techniques in Pattern Recognition"
                John Wiley 1972

Toussant G.T.   "Note on Selection of Independent
                Binary Valued Features for Pattern
                Recognition"
                IEEE Trans. Info Theory, Vol.1,
                IT-17, Sept 1971, p.618

Varghese K.C., Towill D.R., Williams J.H.
                "Selection of Optimum Test
                Frequencies for Fault Diagnosis
                of Analogue Circuits"
                Proc. of the International
                Conference on Technical Diagnostic
                Prague, Czechoslovakia. Aug 1977

Varghese K.C.   "Optimisation of Feature Selection
                and Fault Diagnosis in Analogue
                Systems"
                M.Eng. Dissertation, UWIST Cardiff
                U.K. Jan. 1977

CHAPTER 7
An Analytical Method
of Fault Location

7.1.   Introduction

The methods of feature selection considered have required
large simulation studies and basically heuristic
techniques to generate an optimum sub-set of features.
It would be attractive if the transfer function of the
SUT was able to furnish the information required to
select an optimum set of features and a set of
corresponding unique response patterns.
This approach will enable faults in system parameters to
be isolated. Whether or not the diagnosis gets down to
component level will depend upon the system. Each system
parameter will be, in general, a non-linear function of
many component values, and not readily tractable.
The approach,for convenience,depends on the use of the
frequency domain and in particular the amplitude ratio of
the system. For completeness we briefly consider the
frequency domain theory.

The use of frequency domain test data for fault location in failed systems is well established. Such tests can lead to the optimisation of the number of access points and so to a more robust product (GARZIA,1971). Test frequencies are chosen and using a mathematical model or breadboard simulation a recognition matrix is established representing the fault signatures of the system (VARGHESE,TOWILL,WILLIAMS,1979). Subsequently the fault signature of the system under test is compared with the entries in the recognition matrix and the fault located (VARGHESE,TOWILL,WILLIAMS,1975). The general procedure can be classified as a pattern recognition approach with two practical problems: (i) how to select the test frequencies, (ii) how is the test data used in conjunction with the recognition matrix to identify the fault. Recent work of the Dynamic Analysis Group has concentrated on computer algorithms to establish automatic test program generation (VARGHESE, TOWILL, WILLIAMS,1978). This approach requires the generation of fault signatures having a large number of elements covering the dynamic range of the system under test. The recognition matrix is optimised by comparing each fault signature with its neighbour and discarding those that furnish little or no additional "information". This approach,which was considered in chapter 6, whilst feasible, is heuristic and requires a great amount of computational effort for practical systems. This chapter develops an analytical method of establishing the optimum test frequencies and fault patterns associated with particular system parameters.Using the transfer function representation of the system the square of the amplitude ratio is used. This "power" measurement enables the sensitivity functions to be conveniently established.

7.2.   Test Domain

7.2.1. Introduction

The relationship between input/output of a linear time invariant dynamic system is given by its transfer function, H(s). Where s is the Laplace operator

$$H(s) = \frac{\sum_{i=0}^{p} a_i(s)^i}{\sum_{i=0}^{q} b_i(s)^i} \qquad 7.1$$

The frequency response expression is obtained by substituting jw for s, where $j = \sqrt{-1}$, w=frequency (rad/sec)

$$H(jw) = \frac{\sum_{i=0}^{p} a_i(jw)^i}{\sum_{i=0}^{q} b_i(jw)^i} \qquad 7.2$$

Thus H(jw) is a complex function of w

$$H(jw) = A(w) + jB(w) \qquad 7.3$$

$$[A(w), B(w)\ \text{real}]$$

In engineering terms this complex function is investigated by considering amplitude and phase relative to the input sinusoidal stimulus.

$$|H(jw)| = \sqrt{\left[A^2(w) + B^2(w) \right]}$$

$$\underline{H(jw)} = \tan^{-1}\left[\frac{B(w)}{A(w)}\right] \qquad 7.4$$

These measures are long established and are fundamental in both design and testing of dynamic systems. However, they are not convenient expressions in examining their sensitivity to parameter changes which is fundamental in any diagnostic analysis. It is more mathematically tractable to consider amplitude squared $|H(jw)|^2$. This obviously removes the inconvenient square root, but in addition it emphasises any resonances present.

7.2.2. System Representation

Control engineers have long represented dynamic systems
as consisting of first and second order elements. This
representation is more useful in estimating and
understanding system performance characteristics

$$H(s) = \frac{k\Pi_i(1+\tau_i s)\ \Pi_j(1+\frac{2\zeta_j}{w_{n_j}}s+\frac{s^2}{w_{n_j}^2})}{\Pi_k(1+\tau_k s)\ \Pi_\ell(1+\frac{2\zeta_\ell}{w_{n_\ell}}s+\frac{s^2}{w_{n_\ell}^2})} \qquad 7.5$$

and leads to methodologies enabling the system to be
controlled or tuned using compensating elements.
Thus

$$|H(jw)|^2 = \frac{k^2\Pi_i(1+\tau_i^2 w^2)\ \Pi_j((1-\frac{w^2}{w_{n_j}^2})^2+4\zeta_j^2\frac{w^2}{w_{n_j}^2})}{\Pi_k(1+\tau_k^2 w^2)\ \Pi_\ell((1-\frac{w^2}{w_{n_\ell}^2})^2+4\zeta_j^2\frac{w^2}{w_{n_\ell}^2})} \qquad 7.6$$

represents the analytic function that will be used to
develop the fault location technique and is the form of
actual measurement recommended in this context.

7.3.   Sensitivity Functions
7.3.1. Introduction

The expression given by (7.6) has two basic factors,
namely of first and second order either in the numerator
or denominator. The second order terms have two
parameters ζ and ω_n . Hence the three parameter types of
interest are τ, ζ and ω_n. The sensitivity functions
developed will carry a positive or negative sign
dependent upon whether the parameter in question appears

in the numerator or denominator respectively.
The sensitivity functions are obtained by differentiating
expression (7.6) with respect to the considered parameter
and then normalising with reference to $|H(jw)|^2$ and the
parameter, the general form being

$$S_p = \pm \frac{\delta |H(jw)|^2 / |H(jw)|^2}{\delta p / p} \qquad 7.7$$

S_τ , S_ζ , S_{ω_n} representing the normalised sensitivity
functions with respect to τ , ζ and ω_n .

Table 1 gives the expressions for the three sensitivity
functions.

Parameter	Sp	
		+ for numerator term
		− for denominator term
τ	$\pm \dfrac{2\omega^2\tau^2}{1 + \omega^2\tau^2}$	
ζ	$\pm \dfrac{8\zeta^2\alpha^2}{\left|(1-\alpha^2)^2 +4\zeta^2\alpha^2\right|}$	$\alpha = (\dfrac{\omega}{\omega_n})$ i.e. normalised
ω_n	$\pm \dfrac{4\alpha^2\left|(1-\alpha^2)-2\zeta^2\right|}{\left|(1-\alpha^2)^2 +4\zeta^2\alpha^2\right|}$	

Table 7.1. H(jw) Sensitivity Functions

These sensitivity functions are shown in their normalised
frequency form in figures 7.1a, b and c. For τ the
normalised frequency is $\omega\tau$ and for ζ and ω_n is (ω/ω_n) .
It will be seen that both S_ζ and S_{ω_n} are dependent upon ζ
and the figures show plots for various values of ζ.

114

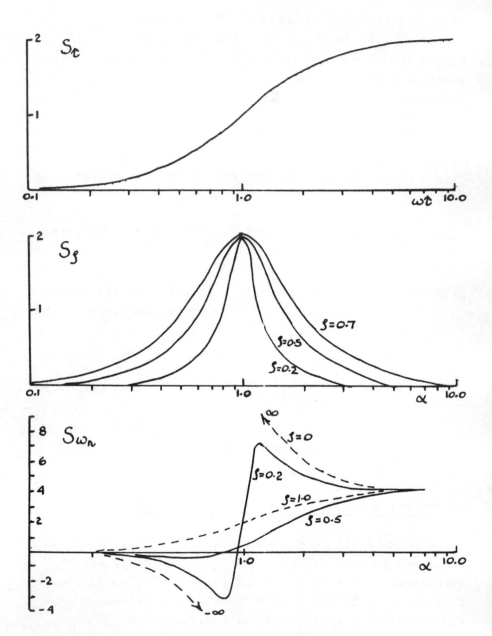

FIG.7.1. Normalised Sensitivity Functions of Table 7.1.

7.3.2. Test frequencies

Having established the sensitivity functions, these can be used to indicate the optimum set of test frequencies to be used in the fault location scheme. This requires the maximum values of the sensitivity functions given in table 7.1, with respect to the normalised frequencies.

Parameter	Sp Maximum	Frequency for Sp max.
τ	± 2	$\omega = \infty$
ζ	± 2	$\alpha = 1$ (ie $\omega = \omega_n$)
$\omega_n \ (\zeta < \frac{1}{\sqrt{2}})$	$\pm 2 - \dfrac{1}{\zeta \sqrt{(1-\zeta^2)}}$	$\alpha^2 = \dfrac{1+2\zeta\sqrt{(1-\zeta^2)}}{1-2\zeta^2}$

Table 7.2. Maximum Values of the Sensitivity Functions

Of interest is the minimum value of S_{ω_n} given by

$$\pm 2 + \frac{1}{\zeta\sqrt{(1-\zeta^2)}} \qquad\qquad \alpha^2 = \frac{1-2\zeta\sqrt{(1-\zeta^2)}}{1-2\zeta^2}$$

It will be seen later that this minimum value helps in identifying ω_n as the likely fault. Figure 7.2 shows the value of α to give the maximum and minimum values of S_{ω_n} for given values of ζ.

7.4. Test measurements

Given the system's transfer function H(s) the nominal response $|H(jw)|^2_N$ can be established. For the system under test under fault conditions the deviations can be formed

116

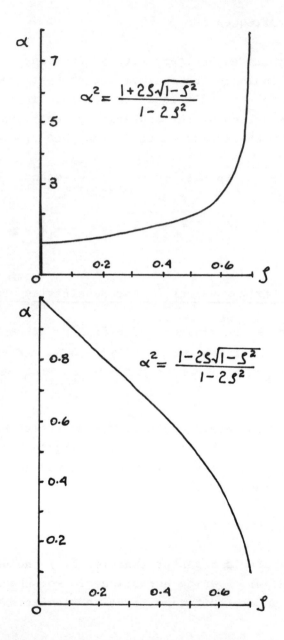

FIG.7.2. Maximum/minimum Values v ζ for S_{ω_n}

$$\Delta^* = |H(jw)|_F^2 - |H(jw)|_N^2 \qquad 7.8$$

This deviation is normalised with reference to $|H(jw)|_N^2$ to give the normalised measure

$$\Delta = \frac{|H(jw)|_F^2 - |H(jw)|_N^2}{|H(jw)|_N^2} \qquad 7.9$$

or

$$\Delta = \frac{|H(jw)|_F^2}{|H(jw)|_N^2} - 1$$

If we assume single parameter changes then Δ takes three forms regardless whether the first or second order term appears in the numerator or denominator, namely

$$\Delta\tau = \frac{1+\tau_F^2\omega^2}{1+\tau_N^2\omega^2} - 1, \qquad \Delta\zeta = \frac{(1-\alpha^2)^2+4\zeta_F^2\alpha^2}{(1-\alpha^2)^2+4\zeta_N^2\alpha^2} - 1, \qquad \Delta\omega_n = \frac{(1-\alpha_F^2)^2+4\zeta^2\alpha_F^2}{(1-\alpha_N^2)^2+4\zeta^2\alpha_N^2} - 1 \qquad 7.10$$

The sign of Δp depends on whether the parameter has increased or decreased from its nominal value. The expressions given in (7.10) match well with the sensitivity functions given in table 7.1, as will be seen later.

7.5. Fault Location Technique - Examples
7.5.1. Introduction

The fault location technique can now be itemised.
- (i)   Select the test frequencies using table 7.2.
- (ii) Obtain the system under test's normalised deviation of amplitude ratio squared.
- (iii) Identify fault from the pattern generated in (ii).

118

7.5.2. Example 1

Consider
$$H(s) = \frac{1}{1 + \frac{2\zeta}{w_n}s + \frac{s^2}{w_n^2}}$$
7.11

with $\zeta = 0.5$, $\omega_n = 1$

From table 7.2 the test frequencies are at

f1 = 1
f2 = 1.93
(f3 = 0.52 Minimum excursion)

The recognition matrix for this example is shown in table 7.3, using plots given in figure 7.1. Two fundamental points emerge from the patterns.

> (i) If the deviations at every test frequency are of the same sign then the likely fault is ζ. Further, the signs of the deviations indicate whether ζ has increased or decreased

> (ii) If the deviations show sign changes then the likely fault is ω_n. Further, the change from negative to positive deviations for increasing frequency or vice-versa indicates whether ω_n has increased or decreased.

Fault	Test Frequencies (rad/sec)		
	0.5	1.0	2.0
ζ +	$-$	$-M$	$-$
ζ $-$	$+$	$+M$	$+$
ω_n +	$-M$	$+$	$+M$
ω_n $-$	$+M$	$-$	$-M$

M= maximum/minimum values.

Table 7.3.   Recognition Matrix for Example 1 Using
           Sensitivity Plots Shown in Figure 7.1.

For convenience the test frequencies given by table 7.2,
namely 0.52, 1, 1.93 rad/sec have been rounded for
convenience to 0.5, 1 and 2 rads/sec. Indicating the
maximum values in this case is superfluous; however, in
systems with more parameters and thus more test
frequencies their use will give assurance in diagnosing
the likely fault.

Each parameter was then changed by $\pm$ 20% in turn and the
normalised response obtained. For completeness the
responses are shown in figure 7.3 over the dynamic range.
The plots clearly exhibit the same fundamental forms as
the sensitivity plots shown in figure 7.1. Table 7.4
gives the deviations as obtained from the 20% changes.

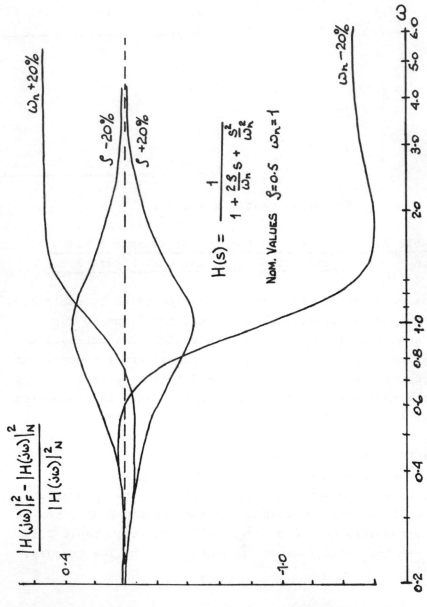

FIG.7.3. Normalised Deviations for Example 1.

Fault	Test Frequencies (rad/sec)		
	0.5	1.0	2.0
ζ +	-0.14	-0.44	-0.14
ζ -	0.11	0.36	0.11
ω_n +	-0.05	0.21	0.54
ω_n -	0.06	-0.88	-1.60

Table 7.4.   Normalised Deviations for Example 1

By comparing tables 7.3 and 7.4 the faults are clearly
identified.   The algorithm required to identify the
likely fault would be very simple.   The recognition
matrix in table 7.3 would be represented by the matrix R,
where

$$R = \begin{bmatrix} -1 & -1 & -1 \\ 1 & 1 & 1 \\ -1 & 1 & 1 \\ 1 & -1 & -1 \end{bmatrix} \qquad 7.12$$

while the test deviation vectors Vi would be represented,
using table 7.4, as

$$V = \begin{bmatrix} -1 & -1 & -1 \\ 1 & 1 & 1 \\ -1 & 1 & 1 \\ 1 & -1 & -1 \end{bmatrix} \qquad 7.13$$

For this example the resulting locating vectors li will
be given by

$$L = VR^T$$

where

$$L = \begin{bmatrix} 3 & -3 & -1 & 1 \\ -3 & 3 & 1 & -1 \\ -1 & 1 & 3 & -3 \\ 1 & -1 & -3 & 3 \end{bmatrix}$$

7.14

the maximum entry in each li indicating the likely fault. In addition, this diagnostic location is emphasised by the most negative entry, indicating the fault in a negative sense.

7.5.3. Example 2

Consider the system

$$H(s) = \frac{1}{(1 + \dfrac{2\zeta_1 s}{w_{n_1}} + \dfrac{s^2}{w_{n_1}^2})(1 + \dfrac{2\zeta_2 s}{w_{n_2}} + \dfrac{s^2}{w_{n_2}^2})}$$

7.15

where $\zeta_1 = 0.5$ $\zeta_2 = 0.2$

ω_{n1} 1.0 ω_{n2} 0.88

These values were chosen to give pairs of complex poles relatively close to each other, which notionally would give problems in identifying which parameter has changed.

Again using table 7.2, the test frequencies are
     0.52, 0.72, 0.89, 1.0, 1.09, 1.93 rad/sec
For convenience these have been rounded to
     0.5, 0.7, 0.9, 1.0, 1.1, 2.0 rad/sec

Each parameter was changed $\pm$ 20% in turn and the resulting patterns are shown in table 7.5.

123

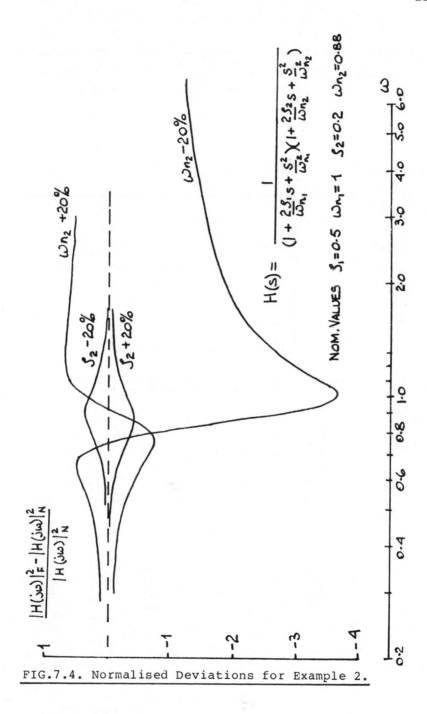

FIG.7.4. Normalised Deviations for Example 2.

$$H(s) = \frac{1}{\left(1 + \frac{2\zeta_1}{\omega_{n_1}}s + \frac{s^2}{\omega_{n_1}^2}\right)\left(1 + \frac{2\zeta_2}{\omega_{n_2}}s + \frac{s^2}{\omega_{n_2}^2}\right)}$$

NOM. VALUES $\zeta_1 = 0.5$ $\omega_{n_1} = 1$ $\zeta_2 = 0.2$ $\omega_{n_2} = 0.88$

Of course the effect of changes in the first factor will be as those shown in figure 7.3. Figure 7.4. shows the results for the second factor. In this case the "maximum" values are indicated.

Fault (20%)	Test Frequencies (rad/sec)					
	0.5	0.7	0.9	1.0	1.1	2.0
ζ_1 +	−	−	−	−M	−	−
ζ_1 −	+	+	+	+M	+	+
ω_{n1}+	(−M)	−	+	+	+	+M
ω_{n1}−	(+M)	−	−	−	−	−M
ζ_2 +	−	−	−M	−	−	−
ζ_2 −	+	+	+M	+	+	+
ω_{n2}+	−	(−M)	−	+	+M	+
ω_{n2}−	(+M)	−	−	−M	−	−

Table 7.5. Recognition Matrix for Example 2

For this example the R matrix is

$$R = \begin{bmatrix} -1 & -1 & -1 & -1 & -1 & -1 \\ 1 & 1 & 1 & 1 & 1 & 1 \\ -1 & -1 & 1 & 1 & 1 & 1 \\ 1 & 1 & -1 & -1 & -1 & -1 \\ -1 & -1 & -1 & -1 & -1 & -1 \\ 1 & 1 & 1 & 1 & 1 & 1 \\ -1 & -1 & 1 & 1 & 1 & 1 \\ 1 & 1 & -1 & -1 & -1 & -1 \end{bmatrix} \qquad 7.16$$

The resulting L matrix is given by

$$
L=
\begin{bmatrix}
6 & -6 & -4 & 4 & 6 & -6 & -4 & 4 \\
-6 & 6 & 4 & -4 & -6 & 6 & 4 & -4 \\
-4 & 4 & 6 & -6 & -4 & 4 & 6 & -6 \\
4 & -4 & -4 & 4 & 4 & -4 & -4 & 4 \\
6 & -6 & -4 & 4 & 6 & -6 & -4 & 4 \\
-6 & 6 & -4 & 4 & 6 & -6 & -4 & 4 \\
0 & 0 & 4 & -4 & 0 & 0 & 4 & -4 \\
4 & -4 & -4 & 4 & 4 & -4 & -4 & 4
\end{bmatrix}
\qquad 7.17
$$

It is immediately obvious that in this case this simple algorithm merely indicates whether either ζ or ω_n has changed. So for systems having multiple parameters of these types the method must include the information as to where the maximums occur. If this is done, then the ambiguity disappears

e.g. ζ_1 all of the same sign with max. @ 1.0 r/s
    ζ_2 all of the same sign with max. @ 0.9 r/s
    ω_{n1} sign change with maximum   @ 2.0 r/s
    ω_{n2} sign change with maximum   @ 1.1 r/s

Of course the more accurate the test frequencies selected the surer will be the diagnostics. In this example the frequencies have been rounded, and the imposed faults are relatively large compared with the sensitivity curves derived using infinitesimal calculus, thus showing the robustness of the method.

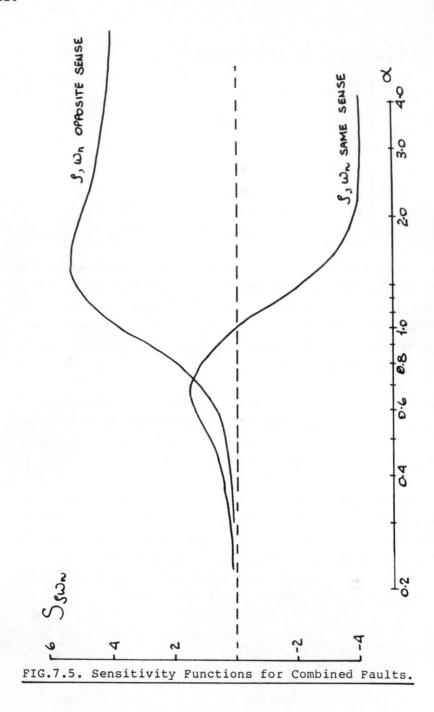

FIG.7.5. Sensitivity Functions for Combined Faults.

7.6.   Multiple Faults

7.6.1. Introduction

So far, we have considered single parameter faults which
in practice are often the case. However, it is possible
for changes in component values to affect more than one
parameter of the dynamic model. In this case the
technique should be capable of detecting changes in both
second order parameters. The sensitivity functions shown
in figure 7.1 can be used to indicate the likely
deviation shapes for given parameter changes. The
various combinations are shown in figure 7.5.

7.6.2. Example 3

Consider the system of Example 1, namely

$$H(s) = \cfrac{1}{1 + \cfrac{2\zeta}{\omega_n} s + \cfrac{s^2}{\omega_n^2}} \qquad\qquad 7.18$$

with $\zeta = 0.5$ and $\omega_n = 1$

The multiple faults imposed are listed in table 7.6

Fault No.	ζ	ω_n
1	+20%	+20%
2	+20%	-20%
3	-20%	-20%
4	-20%	+20%

Table 7.6. Imposed Faults for Example 3

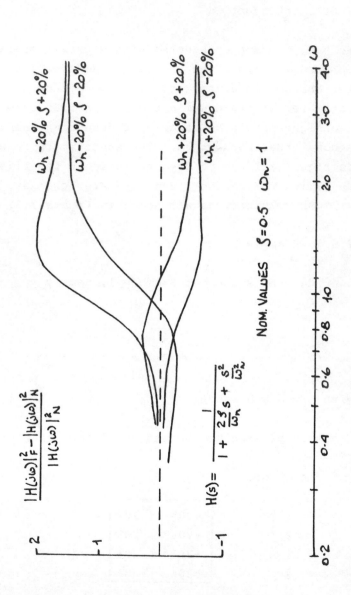

FIG.7.6. Normalised Deviations for Example 3.

The combined sensitivity functions shown in figure 7.5
clearly show the effect of the various combined faults in
ζ and ω_n . The actual plots with the 20% changes are
shown in figure 7.6.
The following expressions give the sensitivity function
for combined faults.

$$\Delta\,{}^{+}_{+} \;=\; -\Delta\,{}^{-}_{-} \;=\; \pm\,\frac{4^{2}(1-\alpha^{2})}{|(1-\alpha^{2})^{2} + 4\zeta^{2}\alpha^{2}|}$$

$$\Delta\,{}^{+}_{-} \;=\; -\Delta\,{}^{-}_{+} \;=\; \pm\,\frac{4\alpha^{2}|4\zeta^{2} - (1-\alpha^{2})|}{|(1-\alpha^{2})^{2} + 4\zeta^{2}\alpha^{2}|}$$

$\Delta\,{}^{+}_{+} = \zeta$ change +ve, ω_n change +ve, etc.

Expressions are +ve. for a numerator term,-ve. for a
denominator term.

The main point emerging from these plots is that the non
zero deviation value for the higher frequencies indicates
an ω_n fault with the enhanced peakiness near the ω_n.
frequency showing the change in ζ .

7.7.   Conclusions

A frequency domain technique has been developed which
enables single parameter changes of a dynamic system to
be identified. The deviation patterns under single fault
conditions are unambiguous even when the system poles (or
zeros) are relatively close together. If multiple
changes exist occurring in different factors of the
transfer function then the level of diagnosability will
depend upon the dynamic separation of the factors. If
the multiple changes exist in the same factor then the
deviation pattern immediately indicates this situation.

REFERENCES

R.F. Garzia   "Fault Isolation Computer Methods"
                NASA Contractor Report CR-1758,
                February 1971, available from NTIS
                Springfield, Virginia 22151, U.S.A.

K.C. Varghese, J.H. Williams, D.R. Towill
                "Simplified ATPG and Analog
                Fault Location via a Clustering
                and Separability Technique"
                IEEE Trans. Circuits and Systems,
                Vol. CAS-26, No.7, p.p. 496-505,
                July 1979

H. Sriyananda, D.R. Towill, J.H. Williams
                "Voting Techniques for Fault
                Diagnosis from Frequency Domain
                Test Data"
                IEEE Trans. Rel., Vol. R-24,
                p.p. 260-267, Oct. 1975
                (Reprinted in IEEE publication
                "The World of Large Scale Systems"
                1982)

K.C. Varghese, J.H. Williams, D.R. Towill
                "Computer Aided Feature Selection
                for Enhanced Analogue System Fault
                Location"
                Patttern Recognition, Vol.10,
                p.p. 265-280, 1978

Index